MASS SPECTROMETRY

Theory and Applications

R. JAYARAM

Microwave Physics Laboratory
Electronics Research Center
NASA
Cambridge, Massachusetts

℗ SPRINGER SCIENCE+BUSINESS MEDIA, LLC 1966

Library of Congress Catalog Card Number 65-25239

ISBN 978-1-4899-5674-3   ISBN 978-1-4899-5672-9 (eBook)
DOI 10.1007/978-1-4899-5672-9

MASS
SPECTROMETRY

To My Mother

Preface

This book on mass spectrometry is the practical result of an intense desire on the author's part to obtain a concise and panoramic view of the recent theoretical and experimental developments in this field. I have always had the feeling that mass spectrometry has been treated as a chemist's subject, and that most of the books written thus far have been directed to the chemist, rather than to the physicist or space-physicist, in spite of the fact that mass-spectrometric techniques have recently been applied extensively in space research, and most of the techniques of mass measurement are outgrowths of contributions made by physicists.

Research on nonmagnetic mass spectrometers received a special stimulus with the advent of space technology and other frontiers of research. The RF mass spectrometers, the time-of-flight mass spectrometers, the quadrupole mass spectrometers, and many other types of nonmagnetic spectrometers stand out as glorious monuments to the painstaking efforts of the scientists involved in their development.

When I completed my study, it appeared worthwhile to share the knowledge with other interested readers. This book is intended to provide an introduction to the various theoretical and experimental techniques in mass spectrometry, with special treatment of the applications of mass spectrometers to the study of ionospheric composition and problems. It is not meant particularly for a specialist in any one field of mass spectrometry, but should prove to be of some help to many students and scientists who wish an introduction to the various research techniques in this field in as short a time as possible.

The book should, in fact, be of interest to both theoretically and experimentally oriented readers, but a good background in differential and integral calculus may be necessary for a facile understanding of the theory underlying the mass-spectrometric techniques presented in the book. Graduate students (in physics, chemistry, or engineering) preparing for a course in mass spectrometry may find the book useful as a text or a reference book, and for chemists, it provides a comprehensive survey of various techniques currently available for mass separation.

I consider it eminently unfair to judge the merit of a work achieved a few years ago in terms of the very recent development in the same field. In fact, today's progress in mass spectrometry is built on the contributions of many investigators in the past who laboriously developed various techniques to the stage where they could be considered reliable. I have tried to review the various applications chronologically, in order to show how techniques underwent radical changes, thereby providing more meaningful structure to the results obtained. Even results which have since become obsolete have been included in some cases for the purpose of tracing historical development.

As many nonmagnetic mass spectrometers as possible have been dealt with in substantial detail. The first chapter provides an introduction to mass spectrometers and spectrographs, particularly the classical magnetic types; the second chapter deals with theoretical and experimental techniques of mass measurement using magnetic mass spectrometers; the third chapter provides a discussion of time-of-flight mass spectrometers. The fourth chapter deals with RF mass spectrometers; the fifth chapter discusses cyclotron resonance instruments as mass analyzers; the sixth chapter deals with the role of mass filters as mass spectrometers; and the last chapter is devoted to the study of applications of nonmagnetic mass spectrometers to upper atmospheric research. Atmospheric composition studies using rocket- and satellite-borne mass spectrometers at various altitude and latitude ranges have been presented. Some of the results of Soviet experiments with RF mass spectrometers have also been detailed. Wherever possible, particular experimental and engineering problems have been discussed. However, no attempt has been made to discuss the most recent results obtained by various investigators concerning the neutral and ionic composition of the atmosphere. This is primarily due to time limitations rather than indifference, and the book will be updated as and when necessary in future.

I am extremely grateful to Dr. Robert D. Stuart, Professor of Electrical Engineering, Northeastern University, for stimulating my interest in mass spectrometry. The present work is due in great part to his influence. It would have been very difficult for me to cope with the time factor without the expert editorial assistance of my friend Denis M. Coffey of Air Force Cambridge Research Laboratories. I am indebted to the Research Administration authorities of Northeastern University, Boston, Massachusetts (where most of the work was done) and the authorities at the Geophysical Institute, University of Alaska, for placing at my disposal many facilities without which it would have been an almost impossible task to bring the material to the publishable stage. Last but not least, I am grateful to the staff of Plenum Press for their patience and expert handling of the whole situation. Permission to use illustrations and other material, the origins of which are indicated in appropriate captions, is gratefully acknowledged. The unsung hero of all this work is Walter Goddard, a draftsman at Northeastern University, who prepared all the diagrams with such effortless ease and elegance, never complaining about time. I acknowledge with gratitude the assistance of all those who have directly or indirectly helped me in the publication of this book.

R. Jayaram

Cambridge, Massachusetts
June, 1966

Contents

Chapter I

Introduction to Mass Spectrometers and Mass Spectrographs

1. MASS SPECTROMETERS AND MASS SPECTROGRAPHS

The term "mass spectrometer" is now usually restricted to an instrument in which separated ion beams are measured electrically. Such instruments are now used in work on isotopic abundances. A "mass spectrograph," on the other hand, is an instrument in which focused ion beams are recorded on a photographic plate. These instruments are used for the exact determination of atomic masses, and such determinations have now become a highly specialized field of investigation. In precision mass spectrographs, it is usual to employ double-focusing methods, whereby a beam of positive ions differing in both mass and kinetic energy is focused to give an exceedingly thin line image on a photographic plate. Both direction and velocity focusing are obtained. The particles of equal e/m (charge-to-mass) ratio, which are brought into focus, have different initial directions and velocities. In mass spectrometers, however, a beam of ions of finite width emerging, say from a slit source, is usually focused by single-focusing to give an image which, even in the ideal case, is as large as the source, and in the practical case is somewhat larger.

There have been many types of mass spectrometers described in the literature and they may be generally classified as being of either the magnetic or the nonmagnetic type. A magnetic-type mass spectrometer is one in which the focusing properties of a magnetic field with respect to charged particles are used directly. Descriptions of various types of nonmagnetic spectrometers and theoretical discussions leading to their design are found extensively in the literature. Kerr (1956) has given an excellent and detailed review of the nonmagnetic types.

A more conventional way of distinguishing among the various types of mass spectrometers is to consider the basic principles involved in each case in effecting a mass separation and analysis. Accordingly, the succeeding chapters will consider the following types: (1) Magnetic sector; (2) time-of-flight; (3) radio-frequency; (4) cyclotron resonance; (5) quadrupole mass filter.

Besides the focusing and analyzing characteristics of mass spectrometers, it is worthwhile to discuss the ion sources, the measurement and detection of ion beams, and the vacuum- and gas-manipulating systems. The remaining sections of this chapter will serve as an introduction to these considerations.

2. ION SOURCES

Electron bombardment sources were gradually developed as vacuum techniques improved. For special investigations, positive ions may be produced from a fused mixture of iron oxide and about 1% of an alkaline-earth or alkali-metal oxide (a Kunsman source) and by the evaporation of ions from a metal filament coated with a salt of the element in question. For this latter source, the work function of the metal composing the filament should exceed the ionization potential of the element. Positive ions may also be produced by allowing a beam of electrons of controlled energy to pass through a gas at low pressure (usually in the range of 10^{-4} to 10^{-7} mm Hg).

However, more complex ion sources are frequently used. The source shown in Fig. I-1, developed by A. O. Nier in 1947, is an example.

The section shown in the figure is perpendicular to the electron beam. In this source, electrons from a tungsten filament are accelerated through a single slit into an enclosed box between P_3 and P_4, which are kept at the same potential as the electron beam. Positive ions passing through the slit in P_4 are accelerated by a moderate electric field between P_4 and J_1 and a larger field between J_2 and J_3. The potentials of J_1 and J_2 are usually equal, as are those of J_3, J_4, and J_5. However, electrode J_4 is divided in two, so that, although its mean potential equals that of J_5, different potentials may be applied to each half to give small sideways deflections to the positive ion beam. Such deflections may be used to compensate for the magnetic deflection produced by the electron-beam collimating field. Electrode J_1 may also be used in the same way.

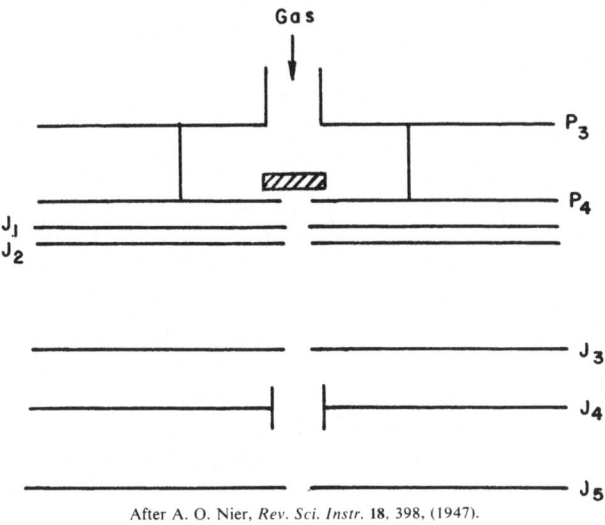

After A. O. Nier, *Rev. Sci. Instr.* **18**, 398, (1947).

Fig. I-1. Nier's ion source.

An expression for the energy distribution of the emitted electrons can be obtained by considering an electron-emitting filament in equilibrium with electrons. If the assumption is made that all the electrons striking the emitter pass into it, so that the reflection coefficient is zero, then, by the principle of detailed balancing, the number of electrons emitted from a unit surface in unit time in a given element of solid angle and with a given range of velocities is equal to the number passing inward with velocities in the diametrically opposite range. The number of electrons with velocities between c and $c + dc$ striking a unit area per second in the equilibrium state is

$$dN = \frac{2\pi m^3}{h^3} \left[\exp\left(-\frac{\phi}{kT}\right) \right] c^3 \left[\exp\left(-\frac{mc^2}{2kT}\right) \right] dc \qquad \text{(I-1)}$$

where h is Planck's constant, k is Boltzmann's constant, ϕ is the thermionic work function, m is the mass of an electron, and T is the temperature. Since the kinetic energy of an electron is given by

$$E = \tfrac{1}{2}mc^2$$

we have

$$dE = mc\,dc$$

Hence

$$dN = \frac{4\pi m}{h^3}\left[\exp\left(-\frac{\phi}{kT}\right)\right]E\left[\exp\left(-\frac{E}{kT}\right)\right]dE \qquad (\text{I-2})$$

where dN is the number of electrons emitted from unit area of the filament per second with kinetic energies between E and $E + dE$.

Oxide-coated filaments or cathodes are very useful if electrons with a low energy spread are required. For investigation of ionization potentials, a strictly monoenergetic beam of electrons, which can be obtained by collimating the beam with AC electric and magnetic fields, is desirable.

3. MEASUREMENT OF ION BEAMS

An ion beam passing through the exit slit in a mass spectrometer is normally collected in a Faraday cylinder. Secondary electrons which may be formed by ion bombardment must be prevented from escaping from the cylinder. Similarly, secondary electrons may also be emitted from the electrode containing the exit slit upon which the ion beam is focused by the magnetic field, and these electrons must be prevented from getting into the Faraday cylinder and hence falsifying the measurements. Since the exit slit and the cylinder are at very nearly the same potential, both effects can be eliminated by placing a wide slit at a negative potential between the exit slit and the cylinder. In a 180° magnetic instrument the strong magnetic field present in the Faraday cylinder region causes secondary electrons to return, in general, to the electrode from which they came.

The positive-ion beam is usually in the range from 10^{-10} to 10^{-15} A, and an electrometer circuit (Fig. I-2) using tubes of very low grid currents is used to measure these ion currents. Tube voltages are kept low to reduce positive ion formation. A typical tetrode electrometer tube, for example, operates with 6 V on the anode and 4 V on the screen grid. The circuit is essentially a Wheatstone bridge, the four arms of which are (a) cathode-anode grid space, (b) cathode-screen grid space, (c) resistance R_1, and (d) resistance R_2. The final output signal (the amplified anode current) ap-

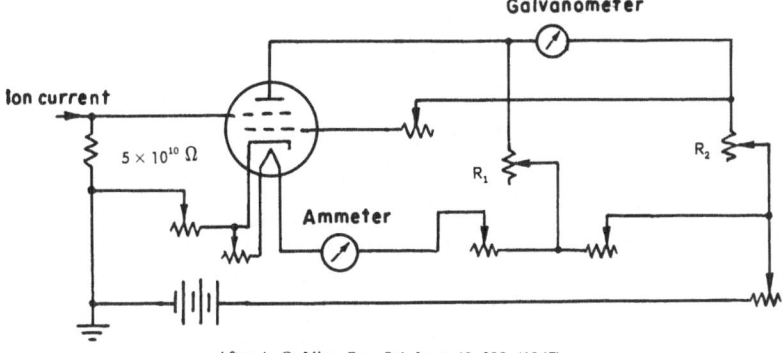

After A. O. Nier, *Rev. Sci. Instr.* **18**, 398, (1947).

Fig. 1-2. Electrometer circuit.

pears in a relatively low resistance circuit, and can be used to operate an automatically recording self-balancing potentiometer.

Electron multipliers and dynamic-condenser electrometers (vibrating-reed electrometers) have also been employed in mass-spectrometric research.

4. PUMPING SYSTEMS

In general, mass spectrometers are operated with continuous pumping and a constant flow of gas passes through the apparatus. The residual gases in mercury diffusion pumps and in hydrocarbon-oil diffusion pumps are detected by the extremely sensitive mass spectrometers in an apparatus totally enclosed in a glass envelope. The mass spectrum observed is known as the background mass spectrum. The pumping rate is generally proportional to $1/\sqrt{M}$, due to molecular flow. The pumping speed in a mass spectrometer depends on both pump speed and the conductance of the appropriate parts of the apparatus.

5. GAS-MANIPULATING SYSTEMS

The gas sample must be admitted to the mass spectrometer ion source through a leak of some kind. If the diameter of the leak is less than the mean free path of the molecules of gas on the high-pressure side of the leak, then the molecular flow of the gas occurs and the rate of admission in molecules per second, if the gas is of molecular weight M, is proportional to $1/\sqrt{M}$ and to the partial pressure of the gas on the high-pressure side of the leak. However, the composition of gas mixture changes with time since lighter gases escape first. But in viscous leaks (long thin capillary), this problem is not present.

6. VACUUM REQUIREMENTS

In analyzing a sample with a mass spectrometer, the substance is converted into ions which are electrically accelerated and passed through some mass-discriminating system (a magnetic field, for instance) and then measured on a collector. To make the measured intensity at a certain mass proportional to the corresponding concentration of ions near the source, it is essential to eliminate or reduce perturbing effects caused by possible collisions of the ions with the gas molecules in the analyzer tube on their way from the ion source to the collector.

The gas in a mass spectrometer tube causes peak-broadening as the ions are diverted from their prescribed trajectories by collisions. Only ions scattered at low angles or near the collector slit can reach the collector. Also the residual gas penetrates the ionization chamber of the source and the ions produced from it are measured together with those from the sample. Hence good vacuum requirements must be satisfied for precision measurements of isotopic masses and their abundances. The vacuum requirements depend on the purpose that is to be served by the mass spectrometer.

In the measurement of isotope ratios the vacuum must be as good as possible. A vacuum of 10^{-8} mm Hg is desirable. However, in the analysis of very small samples, the vacuum requirement may be as low as 10^{-10} mm Hg.

For instruments with very high resolving power, an operating pressure of 10^{-6} mm Hg in the tube is sufficient to prevent intolerable peak-broadening.

In order to keep a firm control over the vacuum level and maintain that level, ionization gauges are utilized as a part of the instrumentation. For the sake of introduction, only one ionization gauge circuit will be discussed here, though many types are in use today.

7. IONIZATION GAUGE CIRCUIT

An ionization gauge provides a convenient means of measuring the high vacuum desired in most mass spectrometers. The gauge usually contains a plate, a grid, and a filament. The grid is positive and the plate negative with respect to the filament, and the electrode voltages are so arranged that the electrons from the filament bombard and ionize some of the gas molecules present in the tube. The resulting positive ions are attracted to a negatively charged plate which is usually cylindrical and encloses the filament and the grid.

The rate of flow of electrons from the plate circuit is used to indicate the gas pressure. In the conventional design, the ratio of plate-to-grid current is proportional to the gas pressure. In the gauge-control circuit shown in Fig. I-3, the stabilized voltages for the gauge are obtained from gas discharge tubes V_1 and V_2. Since the gas pressure is related to the ion current for a certain fixed electron-emission, it is desirable to adjust VR to obtain a filament emission of the correct value.

The control circuit described in Fig. I-3 is useful for pressures above 10^{-5} mm Hg. At this pressure the ion current may be about one microampere. For the accurate measurement of lower pressures, the control circuit must be provided with an amplifier. A circuit incorporating this feature is shown in Fig. I-4.

Tubes V_1 and V_2 constitute a differential amplifier for the gauge ion current. The grid of V_2 is fixed in potential; electrons flowing to the gauge plate through any one of the grid resistors give a reading on the meter M_1 which indicates gas pressure. The amplifier is calibrated by determining from the gauge characteristics what ion current corresponds to the expected pressure. The voltage developed across a resistor can then be calculated and can be applied directly to the grid of V_1 from R_{14}; then the sensitivity control R_8 is adjusted until M_1 reads full scale.

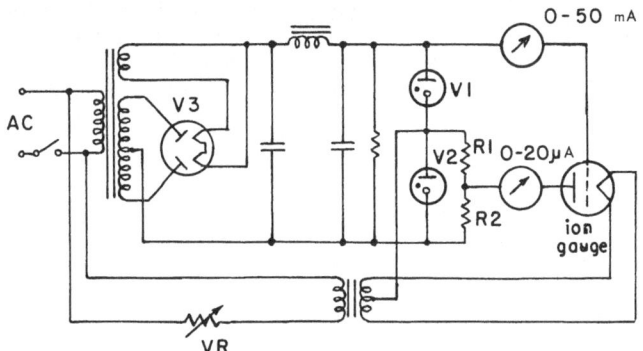

After D. C. Frost, From *Mass Spectrometry* edited by
C. A. McDowell, Copyright © 1963 by McGraw-Hill, Inc.

Fig. 1-3. Ion-gauge control circuit.

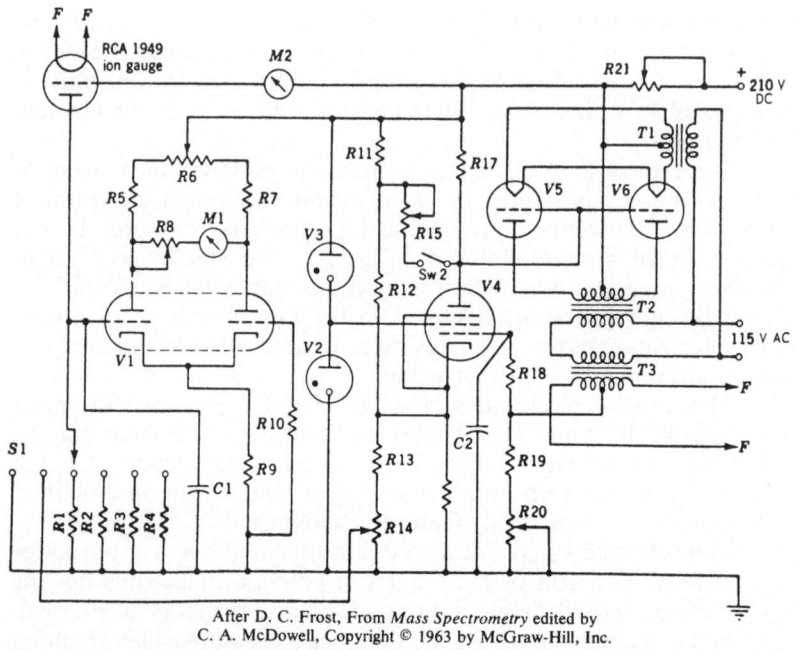

After D. C. Frost, From *Mass Spectrometry* edited by
C. A. McDowell, Copyright © 1963 by McGraw-Hill, Inc.

Fig. 1-4. Stabilized-emission ion-gauge control circuit.
R_1, *1K 1W Aerovox Carbofilm;* R_2, *10K 1W Aerovox Carbofilm;* R_3, *100K
1W Aerovox Carbofilm;* R_4, *1M 1W Aerovox Carbofilm;* R_5, R_7, R_{11}, R_{13},
10K 1W Aerovox Carbofilm; R_6, *5K 1W variable;* R_8, R_{14}, *2K 1W variable;*
R_{10}, R_{17}, R_{18}, *100K 1W;* R_9, *270 Ω 1W;* R_{15}, *10K 1W variable;* R_{16}, *3.9K
1W;* R_{19}, *5.6K 1W;* R_{20}, *3K 1W variable;* R_{21}, *2K 2W WW variable;* C_1,
C_2, *0.02 mF 600-V paper;* S_1, *1P.6 pos. rotary switch;* Sw_2, *1P.2 pos.
on/off switch;* T_1, *transformer, Hammond 165X60;* T_2, *transformer, Hammond 273X60;* T_3, *transformer, Hammond 1144X60;* V_1, *6SC7 tube;*
V_2, V_3, *VR105 tube;* V_4, *6SJ7 tube;* V_5, V_6, *6AS7 tube;* M_1, *0-1 mA meter;*
M_2, *0-10 mA meter.*

    The filament emission is stabilized by means of a servo cir-
cuit. The primaries of T_2 and T_3 are in series with the main supply,
while the secondary of T_3 feeds the filament and the secondary of
T_2, a variable impedance, feeds the tubes V_5 and V_6.
    The filament emission controls the grid of V_4 and, therefore,
the grids of V_5 and V_6. R_{20} sets the emission required, and S_2 and
R_{15} enable a fixed voltage to be applied to the grids of V_5 and V_6
for initial degassing of the gauge.

8. THE WHOLE-NUMBER RULE
AND EXACT ISOTOPIC MASSES

The first mass spectrograph of Aston established the "whole-number rule," according to which the weights of all atoms can be expressed as integral numbers to a considerable degree of accuracy. The relative mass determinations with the earlier mass spectrographs were, however, sufficiently accurate to show slight deviations from whole numbers for certain atoms. J. L. Costa (in Paris, 1925) first attempted accurate measurements of these deviations with an instrument based on the same principles as that of Aston. In 1925, Aston constructed his second mass spectrograph, with which, as a result of various improvements in design, divergences of atomic masses from whole numbers were determined. These measurements were summarized in 1927 in the first "packing fraction" curve. From that time, such data became increasingly significant in relation to nuclear transformations; indeed, K. T. Bainbridge (1933) demonstrated experimentally the equivalence of mass and energy by establishing accurate comparisons of the masses of light particles involved in nuclear disintegrations.

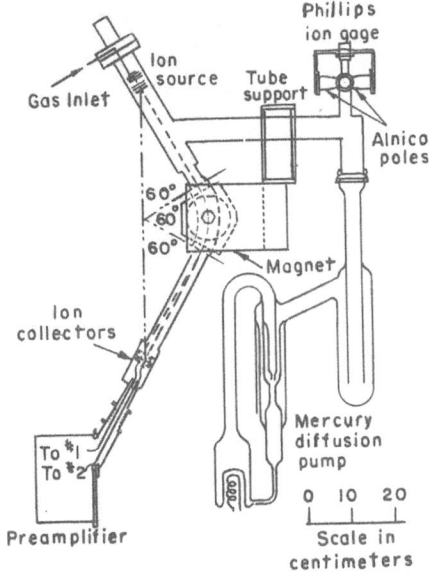

After A. O. Nier, *Rev. Sci. Instr.* 18. 398, (1947).

Fig. I-5. Nier's mass spectrometer.

The determination of the exact relative masses of atoms has become an important field of work, and many instruments of very high precision have been developed for this purpose.

9. SIMPLE MASS SPECTROMETER OF NIER (1947)

A simple mass spectrometer system developed by A. O. Nier in 1947 is presented in Fig. I-5, showing the spectrometer tube, magnet, pressure gauge, and pumping system. This spectrometer system is considered for discussion in order to show how the various experimental techniques are applied within an instrumental framework to achieve mass-discrimination with as high a precision as possible.

The ion source utilized in this system is rather interesting. Electrons are emitted from a tungsten-ribbon filament, pass through an ionizing region, and are caught in a trap. A magnetic field of about 150 G collimates the electron beam. The electron-emission regulator for the spectrometer filament is shown in Fig. I-6.

This circuit also supplies a potential difference for accelerating the electrons. The principle of thermionic control is illustrated in Fig. I-7. T_1 is the transformer supplying power to the load R. The primaries of T_1 and T_2 are connected in series and the AC line voltage is applied to the series combination. Let r_1 and r_2 be the resistances of the secondary and the primary, respectively, of T_1; let

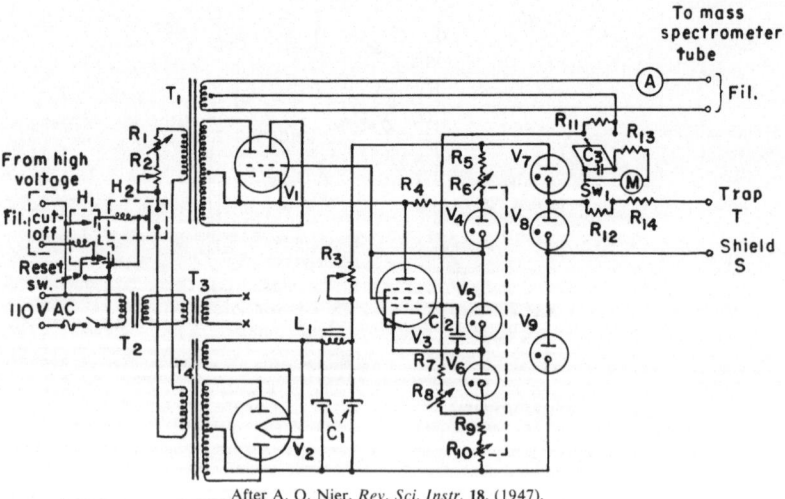

After A. O. Nier, *Rev. Sci. Instr.* **18**, (1947).

Fig. I-6. Electron-emission regulator.

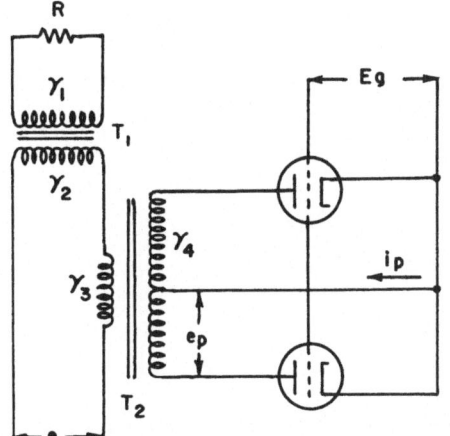

After L. Ridenour and C. S. Lampson, *Rev. Sci. Instr.* 8, 162, (1937).

Fig. I-7. Principle of thermionic control.

m be the primary-to-secondary turn ratio. Let the corresponding quantities for T_2 be r_3 (primary resistance), $2r_4$ (secondary resistance), and $2n$ (turn ratio). Then

$$e_p = ne - \left[\frac{n^2}{m^2}(R + r_1) + n^2(r_2 + r_3) + r_4 \right] i_p \qquad \text{(I-3)}$$

If we write

$$i_p = \frac{e_p + \mu E_g}{R_p}$$

where μ is the voltage amplification factor and R_p the plate resistance of the tube, then

$$i_p = \frac{ne + \mu E_g}{R_p + (n^2/m^2)(R + r_1) + n^2(r_2 + r_3) + r_4} \qquad \text{(I-4)}$$

The instantaneous primary current i is then given by

$$i = ni_p = \frac{n^2 e + n\mu E_g}{R_p + (n^2/m^2)(R + r_1) + n^2(r_2 + r_3) + r_4} \qquad \text{(I-5)}$$

Writing $e = \sqrt{2}\,E \sin \theta$ and $x = -\mu E_g/\sqrt{2}\,nE$, we have

$$i = \sqrt{2}\,A\,(\sin \theta - x)$$

where

$$A = \frac{n^2 E}{R_p + (n^2/m^2)(R + r_1) + n^2(r_2 + r_3) + r_4} \qquad \text{(I-6)}$$

The root mean square primary current is then

$$I = \left[\frac{2}{\pi} \int_{\sin^{-1}x}^{\pi/2} i^2 \, d\theta \right]^{1/2} = A \left[\frac{4}{\pi} \int_{\sin^{-1}x}^{\pi/2} (\sin\theta - x)^2 \, d\theta \right]^{1/2} \qquad (1\text{-}7)$$

Only negative grid biases are considered. Then

$$I = A \cdot F(x) \qquad (1\text{-}8)$$

where

$$F^2(x) = \frac{4}{\pi} \int_{\sin^{-1}x}^{\pi/2} (\sin\theta - x)^2 \, d\theta$$

$$= \frac{4}{\pi} \left[(x^2 + \tfrac{1}{2})\left(\frac{\pi}{2} - \sin^{-1}x\right) - \frac{3x}{2}(1 - x^2)^{1/2} \right] \qquad (1\text{-}9)$$

The average value of the primary current, which is useful in the design of rectifiers employing this control principle, is obtained by simple averaging of i. Let this be I':

$$I' = A \cdot G(x) \qquad (1\text{-}10)$$

where

$$G(x) = \frac{2\sqrt{2}}{\pi} \left[x\left(\sin^{-1}x - \frac{\pi}{2}\right) + (1 - x^2)^{1/2} \right] \qquad (1\text{-}11)$$

and, as before,

$$A = \frac{n^2 E}{R_p + (n^2/m^2)(R + r_1) + n^2(r_2 + r_3) + r_4}$$

The values of the functions $F^2(x)$, $F(x)$, and $G(x)$ for various values of x are listed in Table 1-1.

Table I-1. Corresponding Functions for Various Values of x (Ridenour, Lampson)

x	$F^2(x)$	$F(x)$	$G(x)$
0.00	1.00	1.00	0.90
0.10	0.77	0.88	0.76
0.20	0.58	0.76	0.64
0.30	0.40	0.63	0.52
0.40	0.27	0.52	0.42
0.50	0.17	0.41	0.35
0.60	0.10	0.31	0.23
0.70	0.04	0.20	0.14
0.87	0.004	0.06	0.05
1.00	0.000	0.00	0.00

Nearly all of the past mass spectrometric work has been carried out with magnetic resolution of the ions, but several attempts to employ other principles have been made. In the drift tube method, regular pulses of positive ions are accelerated into a long tube. The lighter ions travel down the tube faster and reach a collector at the end of the tube first. When the collector current is measured as a function of time, a mass spectrum is obtained as each separate pulse arrives at the collector. The measuring circuit is synchronized with the arrival of the pulses, and, accordingly, alternating current methods may be employed, and the mass spectrum can be displayed on a cathode-ray oscilloscope. The resolving power of instruments of this kind made so far has not been very good. In another type of instrument, due to W. H. Bennett (1948), the positive ions are accelerated through a series of grids to which RF potentials are applied. Ions of only one mass arrive at each grid just at the right moment to gain energy, and they are able to pass a retarding potential and reach the collector. These methods eliminate the magnet, but introduce complex problems of electronics. This will be shown in detail in the following chapters.

Chapter II

Magnetic Mass Spectrometers

1. INTRODUCTION

Perhaps the simplest and hence the most thoroughly researched mass spectrometer is the magnetic-sector type. This instrument consists essentially of a uniform magnetic field over a bounded field region or sector. Ions are shot into the sector and only those satisfying particular operating conditions can be collected at the detector. The relative simplicity of this type of instrument is one of its main advantages.

Before going into detailed considerations of focusing properties of electric and magnetic sectors, we find it instructive to introduce some of the terminology involved in mass spectrometry, particularly with regard to resolution.

a. Theoretical Considerations about Resolution

A fundamental requirement of a mass spectrometer is that it resolve an ion beam of mass m from an ion beam of nearly the same mass, $m + \Delta m$. The two ion beams are separated from each other at the resolving slit by a distance given by the mass dispersion D_r of the instrument:

$$D_r = (\text{constant}) \, r_m \frac{\Delta m}{m} \tag{II-1}$$

where r_m is the radius of the ion path in the magnetic field and the constant depends on the geometry of the instrument. If the resolving-slit width is adjusted to be equal to the ion-beam width W_B at the resolving slit, the smallest mass difference Δm which can be completely resolved is determined by

$$D_r = 2W_B$$

or

$$\left(\frac{m}{\Delta m}\right)_{max} = (\text{constant}) \frac{r_m}{W_B} \tag{II-2}$$

where $(m/\Delta m)_{max}$ is called the "resolving power."

In an instrument with minimum aberrations, W_B is equal to $S_o M$, where S_o is the effective source-slit width and M is the magnification. In practical instruments, aberrations which broaden the beam always exist and an increase in r_m above the minimum value is necessary in order to maintain the resolution. Resolution is also critically affected by many other factors. These factors may arise from first-order effects, such as those caused by fringing fields, or from second-order effects, which arise when ions deviate from the path of the central ray, i.e., the optical axis of the instrument.

b. First-Order Effects

The first-order influence of magnetic and electric fields in the analyzer is to bend the ion beam. The required optical properties of the field regions (sectors) result from impulses which are quite small compared to the impulses which cause the optical axis to curve. This is why the optical properties of the instrument are so vulnerable to relative slight imperfections in the fields.

i. Stray Fields. Stray fields refer to any departure from the ideal configuration at the field boundaries. The first-order effect of stray magnetic fields is merely to displace the image position. The quality of the image, however, may not be altered by this effect.

ii. Field Boundaries with Curved Contours. It is found in practice that the fields not only fail to be ideally (abruptly) terminated, but also that the contours of equal field intensity do not remain parallel to the boundary of the field-forming structures. Curvature of the magnetic field boundary can have a pronounced influence upon both the location and the quality of the image.

iii. Space Charge. Space charge is present in both the ion source and the analyzer and its presence always alters the potential and gradient distributions. These alterations by ions of one mass influence the sensitivity of the source to ions of another mass and cause an effect popularly called "interference." This effect limits the linearity with which currents due to two sample components can be superimposed. In the analyzer, space charge causes the ion beam to be larger at the resolving slit, and the resolution suffers as a result.

iv. Surface Charge. Surface charge can completely destroy the optical properties of an instrument. Any high-resistance surface layer which is bombarded by charged particles becomes charged and can cause serious perturbations in the electric field. Care should be taken in designing an instrument to ensure that no surface in the vicinity of the beam is bombarded by the ions.

c. Second-Order Effects

A second-order aberration of importance in the tandem and coincident field-sector instruments is image curvature caused by fringing magnetic fields. For example, if the ions are assumed to travel parallel to the median plane, then for the simple example of a symmetrical magnetic sector, the image-curvature aberration is given by $x = -z_i^2/r_m$, where z_i is the maximum distance from the median plane of ions which are collected. For $z_i = 3 \times 10^{-2}$ in. and $r_m = 10$ in., $x = 0.1 \times 10^{-3}$ in. It is interesting to note that this aberration is absent in the Mattauch–Herzog instrument.

It may be mentioned in passing that the maximum allowable ion current is intimately associated with the resolution required.

d. Time Variance of the Constant Electric and Magnetic Fields

The resolution of a double-focusing instrument, to be discussed later, is not vulnerable to small variations in the accelerating voltage or the ion energy. However, the position of the final image is directly related to the absolute intensity of both the electric and magnetic fields. For all instruments the incremental beam width ΔW_B is given by

$$\Delta W_B = 2r_m \frac{\Delta B}{B} \tag{II-3}$$

where $\Delta B/B$ represents the fractional deviation in the magnetic field intensity.

For a fractional deviation $\Delta E/E$ in electric field intensity, ΔW_B is given by

$$\Delta W_B = r_e \frac{\Delta E}{E} \tag{II-4}$$

For cycloidal instruments, r_m and r_e become focal distances.

2. FOCUSING OF ION BEAMS

a. A General Approach to Magnetic Focusing

As shown in Fig. II-1, a beam of ions diverges from a point source S. The question we must consider is what shape of magnetic field boundary will result in perfect focusing of the beam at a point D. In the following discussion, a symmetrical case is assumed.

The two ion paths, 1 and 2, become circular when they enter the magnetic field region bounded by M. Beam 1, which makes an angle θ with SD, enters the magnetic field at $P(x,y)$ and follows a circular trajectory of radius R. We have

$$x = R \sin \theta \qquad \qquad \text{(II-5)}$$

$$y = (a - x) \tan \theta \qquad \qquad \text{(II-6)}$$

where a is half the distance between S and D (in this symmetrical case). Therefore,

$$\frac{x/R}{y/(a-x)} = \cos \theta = \frac{(R^2 - x^2)^{1/2}}{R} \qquad \qquad \text{(II-7)}$$

and

$$y = \frac{x(a - x)}{(R^2 - x^2)^{1/2}} = f_m(x) \qquad \qquad \text{(II-8)}$$

A field conforming to these boundary conditions is referred to as an "ideal focusing field." The parameters a and R will affect the shape of the ideal field as shown in Fig. II-2.

If R is greater than a it is impossible to obtain the focusing of widely divergent beams.

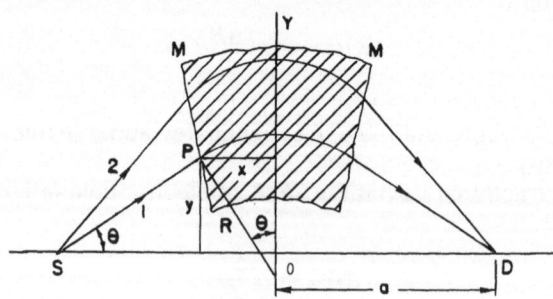

Fig. II-1. Geometry of magnetic focusing (Kervin).

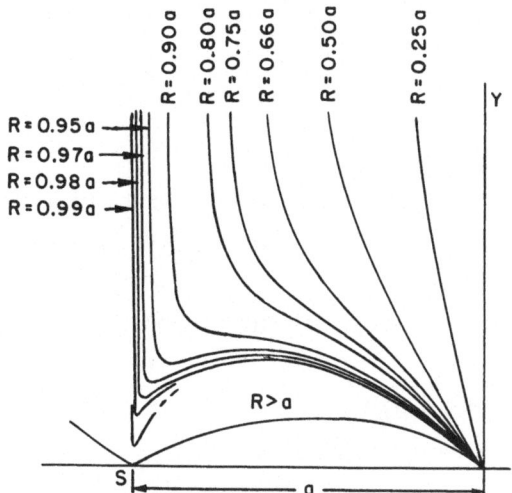

Fig. II-2. Choice of parameters a and R.

i. Image Displacement in Magnetic Sectors. Consider the displacement of the ray of ions shown in Fig. II-3.

After leaving the source S at an angle θ with SD to enter the magnetic field bounded by A_1 (x_1y_1) the ray follows a curved path of radius R and emerges through the boundary A_2 at x_2y_2 to cross the base line at D. The distance SD $(= L)$ may be determined in terms of the ray and field parameters θ, R, x_1, x_2, and a.

From Fig. II-3,

$$L = (a-x_1) + R \sin \theta + R \sin \phi + y_2 \cot \phi \qquad \text{(II-9)}$$

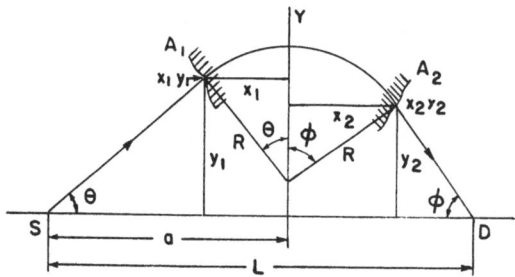

Fig. II-3. Geometry of ion displacement.

For the symmetrical case, $\phi = \theta + \epsilon$, where ϵ is small, so that

$$L = 2a - \frac{(a - x_1)(x_2 - x_1)}{R \sin \theta \cos^2 \theta} \approx 2a \qquad \text{(II-10)}$$

for symmetrical ideal field.

An expression for the displacement of the focusing point D is

$$dL = 2dx_1 + \frac{(a - x_1)(dx_1 - dx_2)}{R \sin \theta \cos^2 \theta} \qquad \text{(II-11)}$$

By suitable approximations and substitutions, one obtains

$$dL = \frac{2n}{[\tan \theta + f_m'(x)]^2}\left[\tan \theta + f_m'(x) + \frac{(a - x)f_m'(x)}{R \sin \theta \cos^2 \theta} \right] \qquad \text{(II-12)}$$

where n is the vertical distance between the curves M and A in the vicinity of either x_1 or x_2 (symmetrical case), and

$$f_m(x) = \frac{x(a - x)}{(R^2 - x^2)^{1/2}} \qquad \text{(II-13)}$$

and $f_m'(x)$ is the derivative of $f_m(x)$ with respect to x.

ii. α-Focusing. One may extend and generalize this relation to the case of a beam of ions of divergence α leaving S at an angle θ with the base line. Substituting $n = f_n(dx)$, dL is then given by

$$dL = \frac{2}{[\tan \theta + f_m'(x)]^2} f_n\left\{ \frac{(a - x)\alpha}{[\tan \theta + f_m'(x)]\cos^2 \theta} \right\}$$

$$\times \left[\tan \theta + f_m'(x) + \frac{(a - x)f_m'(x)}{R \sin \theta \cos^2 \theta} \right] \qquad \text{(II-14)}$$

The aberrations produced by the focusing action of magnetic fields on beams in planes parallel to the pole faces are referred to as "α-aberrations" and the focusing action is referred to as "α-focusing."

The aberration is normally found to be

$$A_n = R\alpha^2 \qquad \text{(II-15)}$$

where R is the radius of curvature.

Figure II-4 depicts cases for $\theta = 30$, 45, and 90°, which are very commonplace.

Of particular interest is the 90° case, for which the two sides of the magnetic field become a single straight line and the ion deviation is 180°. It may also be observed that in circular focusing, perfect direction focusing occurs after the beam has turned through

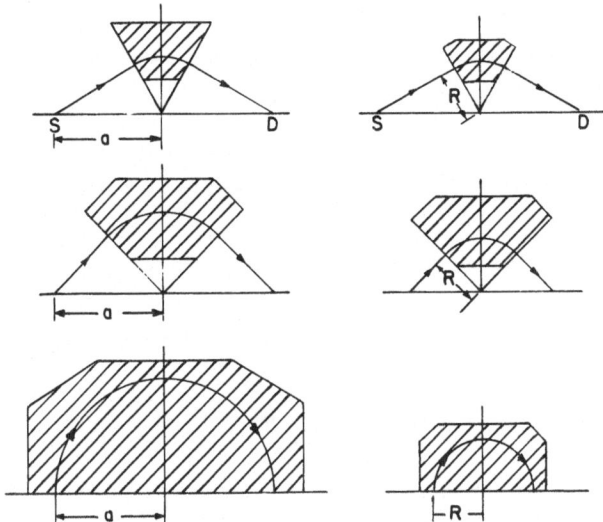

Fig. II-4. Several cases of normal focusing fields.

360°. This property can be fruitfully utilized only if a combination electric and magnetic field is used, since it is necessary to avoid overlapping of the source and the detector. Smith used this property in his instrument, where the ion beam from the source, after a preliminary 180° (orbit 1) turn, is decelerated by a retarding electric field at the pulser slit. The beam then makes a number of circular turns along orbit 2, missing the source and increasing the time dispersion between masses at every turn. Being gradually retarded by suitable application of synchronized retarding pulses, the ion beam follows orbits of smaller and smaller radii and ultimately falls into a detector properly positioned along the path of the nth orbit.

iii. β-Focusing. As opposed to α-focusing, which concerns ion optics of rays in planes parallel to the pole face, the term β-focusing is concerned with the ion optics of rays which move at an angle with respect to these planes. Only the case of homogeneous fields will be discussed here.

Viewed in the plane of the magnetic pole face, most spectrometers have slits which are very narrow. However, the ion-beam divergence α is not negligible. When viewed at right angles to this plane, the beam is usually much narrower at the center of the trajectory, being confined to the pole-gap width at most, as shown in Fig. II-5.

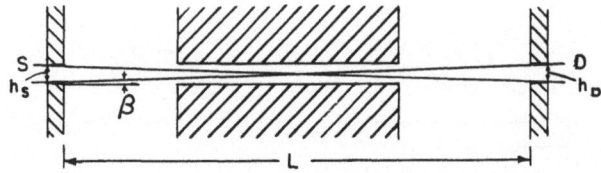

Fig. II-5. Source slit extending right across pole gap width.

The maximum angle which an ion ray can make with the pole face is given by

$$\beta = \frac{h_S + h_D}{2L} \qquad (II-16)$$

where h_S and h_D are the source- and detector-slit lengths, respectively, and L is the length of central ion trajectory.

The ions traveling at an angle with respect to the pole face will be deflected less than those traveling in paths parallel to the pole faces and will arrive at the detector separated by an amount referred to as "β-aberration."

Thus, the variation in radius of curvature is

$$\Delta R = R \frac{\Delta H}{H} \qquad (II-17)$$

where $\Delta H = H - H \cos \beta$. When β is small, $\Delta H \approx H\beta^2/2$ and

$$\Delta R \approx \frac{R\beta^2}{2} \qquad (II-18)$$

The dispersion is approximately $2\Delta R$. As a result, the β-aberration for the normal case is

$$A_{\beta n} = R\beta^2 \qquad (II-19)$$

The β-aberration must be added to the α-aberration and other components in calculating total image width.

iv. Dispersion. A general expression for the dispersion produced by any symmetric approximation to the ideal field is

$$D = 2\Delta R \sin \theta + \frac{n}{[\tan \theta + f'(x)]^2} \left[\tan \theta + f'(x) + \frac{(a - x)f'(x)}{R \sin \theta \cos^2 \theta} \right]$$

$$(II-20)$$

In the normal circle case, D takes the form

$$D_{nc} = 2\Delta R \left[1 + \frac{\Delta R}{R} \tan^2 \theta \right] \frac{M}{\Delta M} \tag{II-21}$$

where M is the ion mass and ΔM is the mass-dispersion.

v. Resolving Power. A good mass spectrometer will effectively accomplish simultaneous separation and sharp focusing of ions of different masses. A measure of the resolving power of such an instrument is the ratio of its dispersion to its bandwidth.

If a spectrometer focuses two ion beams differing in mass by ΔM at two points separated by Δx, then the dispersion is

$$D = \frac{\Delta x}{\Delta M} \tag{II-22}$$

and the resolving power is

$$r.p. = \frac{M}{\Delta M} \tag{II-23}$$

Combining the above two equations,

$$r.p. = \frac{M}{\Delta x} D \tag{II-24}$$

and two ion beams are just resolved when Δx is equal to the beam width.

A generalized expression for the resolving power, taking into account all the intrinsic errors, is

$$p. = \frac{R\left(1 - \dfrac{\tan \alpha}{\tan \gamma}\right)}{(D + I + A + B + E + S_c + F)\left(1 - \dfrac{\tan \alpha}{2 \tan \gamma}\right) - t\left(\dfrac{\tan \alpha}{2}\right) + T\left(1 - \dfrac{\tan \alpha}{\tan \gamma}\right)}$$

$$\tag{II-25}$$

where A is the α-aberration, B is the β-aberration, I is the image of the source slit, E is the energy dispersion (ions in the beam are assumed to be monoenergetic, but in practice they are not; this gives rise to energy dispersion), S_c is the space charge effect, which is directly proportional to beam intensity and inversely to beam width (this effect is usually negligible for currents smaller than 10^{-9} A), F is the fringing-field aberration, D is the detector-slit width, γ is the effect of the slit width and the line of focus (the line of focus of the ion beam does not lie in the plane of the slit, but at an angle γ to it), t is the slit thickness, α is the half-angle of beam divergence, and T is an extra factor added to the total width of the detector if automatic scanning techniques are employed.

b. Focusing Properties of a Homogeneous Magnetic Field

To gain further insight into the subject of magnetic mass spectrometers, one should consider the ion optics and associated theory of a deflection-type mass spectrometer.

Figure II-6 shows a sector of a homogeneous magnetic field. O, I, and C are the positions of the object, image, and center of curvature, respectively; l'_m and l''_m are the object and image distances, respectively, from the boundaries of the magnetic field; b'_m and b''_m are used to describe the object and image widths, respectively, or to measure the displacement of object or image in a direction normal to the central path; a_m is the radius of curvature of the ion beam in the magnetic field; Φ_m is the angle of deflection; and α is the half-angular directional spread of the ions.

Let us assume that ions of mass M_o and velocity v_o emerge from the object point O with half-angular width α in the plane of the paper. An ion with median direction, after traversing the distance l'_m, enters normally the magnetic field, where it is constrained to follow a circular path of radius a_m. After deflection through the angle Φ_m, it emerges from the field normally and proceeds to the image point I, where the ion beam which diverged from O now converges.

Herzog likened this arrangement to the optical combination of a prism and a cylindrical lens, and demonstrated that its focal length is given by

$$f_m = \frac{a_m}{\sin \Phi_m} \tag{II-26}$$

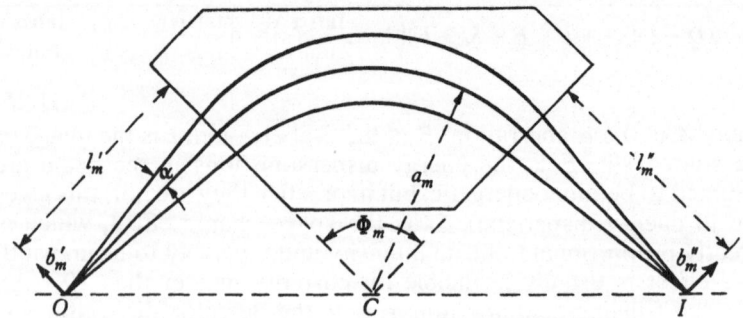

From *Mass Spectroscopy* (Henry E. Duckworth), Cambridge University Press, 1958.

Fig. II-6. Focusing properties of a homogeneous magnetic field.

The object and image distances l'_m and l''_m are related through the equation

$$(l'_m - g_m)(l''_m - g_m) = f_m^2 \qquad (\text{II-27})$$

where $g_m = f_m \cos \Phi_m$ is the distance from the boundary of the field to the principal focus.

The displacement b''_m of the image, corresponding to a displacement b'_m of the object, allowing for a change of ionic mass to $M = M_o (1+\gamma)$ and of ion velocity to $v = v_o (1+\beta)$, where both γ and β are small quantities, is given by

$$b''_m = a_m (\beta + \gamma) \left(1 + \frac{f_m}{l'_m - g_m}\right) - b'_m \frac{f_m}{l'_m - g_m} \qquad (\text{II-28})$$

When $\gamma, \beta \ll 1$,

$$\left|\frac{b''_m}{b'_m}\right| = \frac{f_m}{l'_m - g_m} \qquad (\text{II-29})$$

which has unit value for all symmetrical arrangements ($l'_m = l''_m$).

If $S_o = 2b'_m$ is the width of the object slit located at O, the images corresponding to two mass groups M_o and $M = M_o(1 + \gamma)$ are just resolved when their centers are separated by the distance $2b''_m = S_a f_m/(l'_m - g_m)$.

For the common case of a monoenergetic ion beam, where $\gamma + 2\beta = 0$, the resolution is obtained from (II-28) in the form

$$\frac{\Delta M}{M} = \frac{2S_o}{a_m} \frac{f_m}{(l'_m - g_m)[1 + f_m/(l'_m - g_m)]} \qquad (\text{II-30})$$

For a symmetrical arrangement,

$$\frac{\Delta M}{M} = \frac{S_o}{a_m} \qquad (\text{II-31})$$

In the latter case, the resolution is improved by a factor of two if the monoenergetic ion beam is replaced by one in which all the ions possess the same velocity.

The normal procedure is to replace the image point I with a slit width S_i, followed by a sensitive electrical detector. In this case, the resolution for a symmetrical arrangement becomes $(S_o + S_i)/a_m$, rather than simply S_o/a_m. For maximum sensitivity this image slit should not be smaller than the actual width of the image. The dispersion D, measured in the plane normal to the central beam, is given by $D = (a_m/100)$ cm for 1% mass difference. This is a first-order theory since α and β have been assumed to be very small.

Theories for oblique incidence have been developed by Herzog (1934) and Cartan (1937).

c. Further Discussion on Magnetic Focusing of Ion Beams

From typical ion sources a beam of ions, all of which have very nearly the same kinetic energy, can be produced. In the presence of a magnetic field at right angles to the direction of motion of the ion beam, each ion experiences a force at right angles to both its direction of motion and the magnetic field direction. An ion of charge e moving with a velocity of u cm/sec experiences a force of Heu/c dynes in a magnetic field of H oersteds, where H is at right angles to u. This force must be balanced by a centrifugal force. The ion will therefore describe a circular path of radius r cm, satisfying the condition

$$\frac{mu^2}{r} = \frac{Heu}{c} \tag{II-32}$$

If V is the potential of the ion, $\frac{1}{2} mu^2 = eV$, then

$$r^2 = \frac{2VC^2}{H^2}\left(\frac{m}{e}\right) \tag{II-33}$$

A slightly diverging beam of ions is brought to a focus after a deflection of 180° in a magnetic field. Ions of different mass and equal charge can be separated as shown in Fig. II-7.

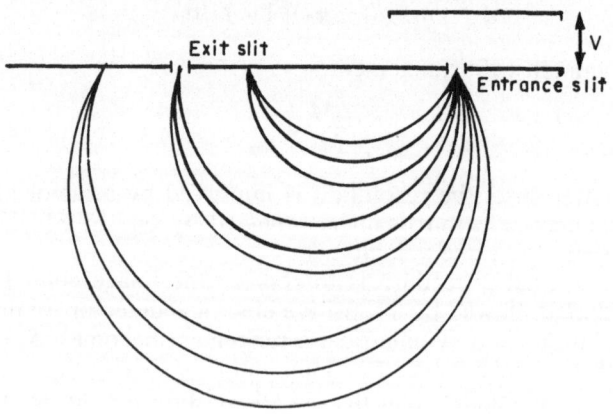

Fig. II-7. Magnetic focusing of ion beams.

The different ion beams may be focused on an exit slit by adjusting either V or H. This arrangement is known as 180° mass spectrometer. However, the focusing of a divergent beam by deflection through 180° in a magnetic field is not perfect. The geometry of magnetic focusing is illustrated in Fig. II-8. Let ABC represent a central ray which intersects the line AC at right angles and let ADE be a divergent ray which makes an angle α with the central ray. ADE meets the line AC at E. The error in focusing is CE, and may be described as *spherical aberration*. If AF is the diameter of a circle of which ADE is part, then $A\hat{E}F$ is a right angle. Also $E\hat{A}F = \alpha$, and therefore

$$AE = 2r \cos \alpha$$

and

$$CE = 2r(1 - \cos \alpha)$$

If α is small, then

$$CE \approx r\alpha^2 \tag{II-34}$$

The ion beam entering the magnetic field is defined by an entrance slit of finite size. For perfect first-order focusing, an image of the entrance slit would be produced on the line AEC equal in width to the entrance slit.

A further error in focusing arises from nonuniformity of the magnetic field. The value of H is greater near the poles than it is in the center of the gap. Thus, ions at the edge of the beam are

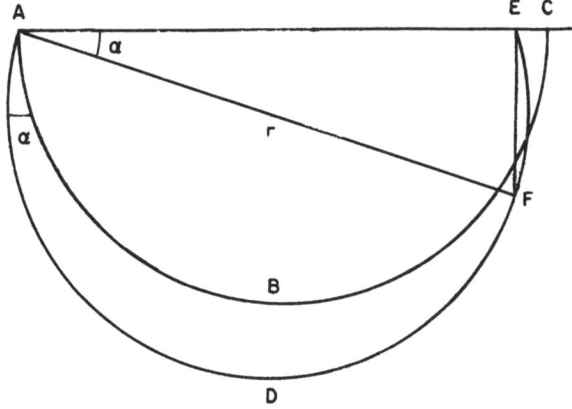

Fig. II-8. Geometry of magnetic focusing.

deflected more strongly by the magnetic field. Additional broadening of the image width can be caused by instability of V and H. Electromagnets greatly reduce inhomogeneity of magnetic field. A. J. B. Robertson used a magnet with semicircular pole pieces, and placed the ion source just outside these in the stray-field region. The magnet gap was just wide enough (2.7 cm) for the magnetic analyzer and its glass envelope. A. O. Nier in 1940 described an instrument in which the ion beam is deflected through only 60° by a sector-shaped magnet as shown in Fig. II-9.

The central ray from the entrance slit falls normally upon the magnetic field boundary and the ion beam can then be focused upon the exit slit provided that the two slits and the apex of the wedge-shaped field all lie upon a straight line. With this type of focusing, the ion source is quite outside the magnetic field and a smaller gap between the pole pieces can be used. Sufficient magnetic field strength can be generated by a moderately light magnet. The spherical aberration with wedge-shaped magnetic field is again $r\alpha^2$. Both electrical and magnetic scanning methods are used.

d. Direction Focusing in Radial Electrostatic Fields

The focusing properties of radial electrostatic fields can be analyzed using Fig. II-6 and replacing subscript m by subscript e. This is sufficient for the purpose of analysis. A field of angular extent $\pi/\sqrt{2}$ corresponds to the field of a semicircular magnet, in that a beam of ions diverging from a point at the entrance boundary

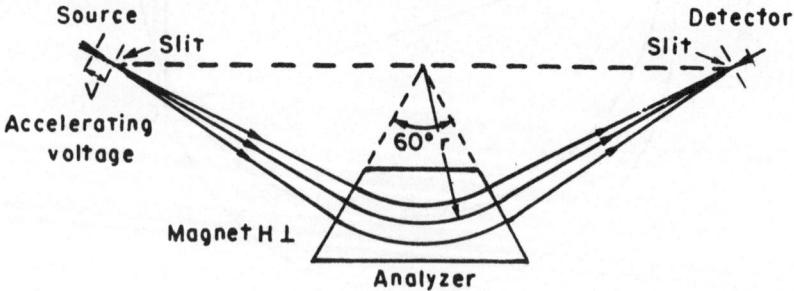

Fig. II-9. Focusing of a divergent ion beam by a wedge-shaped magnetic field.

is brought to a focus at the exit boundary. Herzog (1934) developed the general focusing equations

$$(l_e' - g_e)(l_e'' - g_e) = f_e^2$$

where

$$f_e = \frac{a_e}{\sqrt{2} \sin \sqrt{2}\, \Phi_e} \qquad (\text{II-35})$$

and

$$g_e = f_e \cos \sqrt{2}\, \Phi_e$$

The image displacement b_e'' is given by

$$b_e'' = a_e \left(\beta + \frac{\gamma}{2} \right)\left(1 + \frac{f_e}{l_e' - g_e} \right) - b_e' \frac{f_e}{l_e' - g_e} \qquad (\text{II-36})$$

The magnification, resolution, and dispersion may then be calculated in the same way as in the case of a homogeneous magnetic field.

In the case of a monoenergetic ion beam for which $\beta = -\gamma/2$, all ions regardless of their mass are focused at the same point. The radial electrostatic analyzer is thus an energy selector. Its chief use in mass spectroscopy is in linear combination with a magnetic analyzer to form a double-focusing arrangement.

Warren and Powell at the University of Wisconsin designed an electrostatic analyzer as early as 1947 for the selection of homogeneous ion beams. The analyzer is built with a high-pressure electrostatic generator. The proton beam from the generator is deflected through a 90° arc of 40 in. radius by means of a radial electric field between curved metal plates $5/16$ in. apart. Suitably disposed slits select a beam which has a very small energy spread. The apparatus is operated with energy resolutions up to 5000, and preliminary tests are made using the 985 and 1020 kV γ-ray resonances in the reaction $Al^{27} + H^1 \longrightarrow Si^{28} + h\nu$.

e. Double Focusing

i. Double Focusing in Consecutive Field Combinations. The β term in the expression for b'' is given previously in the form

$$b_m'' = a_m (\beta + \gamma)\left(1 + \frac{f_m}{l_m' - g_m} \right) - b_m' \frac{f_m}{l_m' - g_m} \qquad (\text{II-37})$$

$$b_m'' = - b_m' \frac{f_m}{l_m' - g_m} \qquad (\text{II-38})$$

when γ, $\beta \ll 1$. The β term gives the velocity dispersion associated with that particular field and makes it possible to plan a two-field arrangement in which the velocity dispersion produced in one will be counterbalanced by that produced in the other. If measures are simultaneously taken to ensure that direction focusing is achieved by the combination, the final image will be independent of both velocity and direction spread among the ions emerging from the object slit. This combination will therefore be double-focusing.

The most commonly employed double-focusing arrangement is the one in which the image formed by a radial electrostatic field serves as the object for a following magnetic field. Thus from equations (II-36) and (II-37) the final image displacement is given by

$$B_m'' = a_m(\beta + \gamma)\left(1 + \frac{f_m}{l_m' - g_m} \right) - \frac{f_m}{l_m' - g_m}$$
$$\times \left[a_e \left(\beta + \frac{\gamma}{2} \right) \left(1 + \frac{f_e}{l_e' - g_e} \right) - b_e' \frac{f_e}{l_e' - g_e} \right] \qquad \text{(II-39)}$$

Velocity focusing will occur when the coefficient of β vanishes; that is, when

$$a_m \left(\frac{l_m' - g_m}{f_m} + 1 \right) - a_e \left(1 + \frac{f_e}{l_e' - g_e} \right) = 0 \qquad \text{(II-40)}$$

in which case equation (II-39) reduces to a simpler form from which the resolution R and the dispersion D may be computed. They turn out to be

$$R = \frac{\Delta M}{M} = \frac{2S_o}{a_e} \frac{f_e/(l_e' - g_e)}{1 + f_e/(l_e' - g_e)} \qquad \text{(II-41)}$$

$$D = \frac{a_m}{200} \left(1 + \frac{f_m}{l_m' - g_m} \right) \text{ [cm per 1\% mass difference]} \qquad \text{(II-42)}$$

The resolution is a function of the constants of the electrostatic analyzer, while dispersion is a function only of the constants of the magnetic analyzer.

ii. Focusing in Crossed Electric and Magnetic Fields. The first double-focusing scheme (Dempster, 1929) actually consisted of a deflection through $\pi/\sqrt{2}$ radians in crossed magnetic and radial electrostatic fields. The object and image are located at the entrance and exit boundaries, respectively.

Some years later, Bleakney and Hipple (1938) showed that the combination of crossed uniform magnetic and electric fields possesses the property of perfect double focusing in the plane normal to the magnetic field.

The negative ions with specific charge, emerging from the object slit O, regardless of their direction or velocity in the plane perpendicular to the magnetic field, will pass through the image point I.

The ions follow trochoidal paths, the nature of which depends on the initial conditions. Curtate and prolate orbits are described by the ions. The ions following the former type of path emerge obliquely from the object slit, while those following the latter path emerge more or less normally. The ordinary cycloidal path from O to I will be traced by ions emerging normally with zero velocity.

The radius a of the circle generating the cycloid is given by

$$a = \frac{EMc^2}{eH^2} \qquad \text{(II-43)}$$

where E and H are the electric and magnetic field strengths, respectively. The distance between O and I is given by

$$b = 2\pi a = \frac{2\pi EMc^2}{eH^2} \qquad \text{(II-44)}$$

Therefore, b is proportional to (M/e), a linear mass scale. The resolution is given by

$$R = \frac{\Delta M}{M} = \frac{\Delta b}{b} = \frac{S_o}{b} \qquad \text{(II-45)}$$

where S_o is the object slit width. The dispersion is given by

$$D = \frac{2\pi Ec^2 \Delta M}{eH^2} \text{ cm} \qquad \text{(II-46)}$$

for 1% mass difference and $\Delta M/M = 0.01$.

3. DOUBLE-FOCUSING INSTRUMENTS

a. Introduction

The optical properties of magnetic and electric sectors give what is called "*first-order direction focusing or α-focusing.*" If, in addition, the position of a line image is first-order independent of velocity deviations (β), the instrument is said to be *double-focusing*, for which both electric and magnetic fields are essential. Combination of these two fields can be accomplished in a variety of ways. One of the simplest is to place in sequence an electric sector and a magnetic sector (tandem sectors). In this case, the image of the source formed by the first sector becomes the object for the second sector. In any event, the mass separation takes place in the magnetic field only. The addition of the electric sector improves the definition of the focus by eliminating the first-order velocity aberrations. An alternate way of accomplishing double focusing is to superimpose an electric and a magnetic sector (coincident field). A third way is to superimpose uniform electric and magnetic fields, with internal object and image points, in which ion paths are cycloidal (coincident field cycloid).

b. Dempster's Mass Spectrograph (Dempster, 1935)

As shown in Fig. II-10, this instrument consists of a 90° radial electrostatic analyzer followed by a semicircular magnetic analyzer. The principal slit S_2 is located at a distance $l'_e = 0.150a_e$ in front of the effective boundary of the electrostatic field. The image formed by the electrostatic analyzer is at a distance $l''_e = 0.609a_e$ beyond it, and is located at the effective boundary of the magnetic field, thus making $l'_m = 0$. The final image is formed at the exit boundary of the magnetic field ($l''_m = 0$), at the lower surface of the photographic plate. Velocity focusing occurs for one radius of curvature

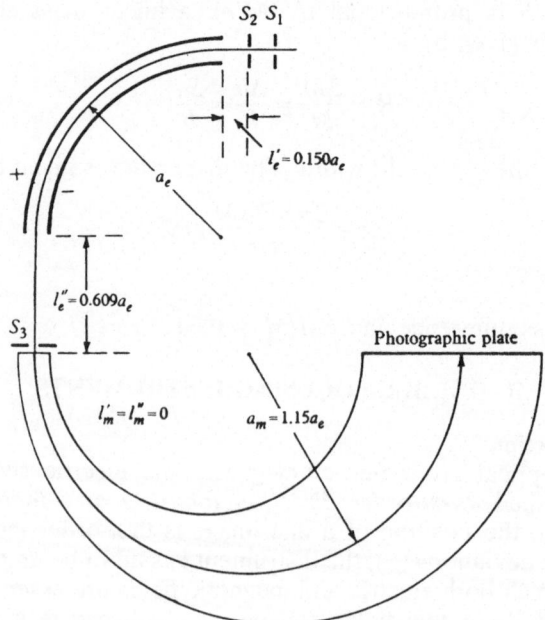

From *Mass Spectroscopy* (H. E. Duckworth), Cambridge University Press, 1958.

Fig. II-10. Dempster's double-focusing mass spectrograph.

in the magnetic field, namely $a_m = 1.15a_e$. From equations (II-41) and (II-42) one may obtain the value of resolution, dispersion, and magnification in the following form:

$$\text{Resolution} = \frac{\Delta M}{M} = \frac{S}{0.88a_e} \tag{II-47}$$

$$\text{Dispersion} = D = 0.01a_m \text{ cm per 1\% mass difference} \tag{II-48}$$

$$\text{Magnification} = \frac{b_m''}{b_e'} = 1.29 \tag{II-49}$$

S_3 guarantees that only those ions whose energies lie within the narrow range permitted by the Herzog theory enter the magnetic analyzer.

In Dempster's original instrument, $a_e = 8.5$ cm and $a_m = 9.8$ cm, corresponding to a resolution of 1/3000 for a 0.0025-cm principal slit, and a dispersion of 0.098 cm/1% mass difference.

c. Wien Velocity Filter

The velocity filter consisting of crossed electric and magnetic fields, which had been used in Bainbridge's first mass spectrograph, was found by Herzog (1934) to possess a direction-focusing property. The focusing condition is expressed as follows:

$$(l_w' - g_w)(l_w'' - g_w) = f_w^2 \tag{II-50}$$

with

$$f_w = \frac{a_m}{\sin(L/a_m)} \qquad g_w = f_w \cos(L/a_m) \tag{II-50}$$

where a_m is the radius of curvature which would result if the magnetic field of the filter were acting alone, and L is the length of the filter. The image displacement is given by

$$b_w'' = -a_m \beta \left[1 + \frac{f_w}{l_w' - g_w} \right] - b_w' \frac{f_w}{l_w' - g_w} \tag{II-51}$$

Therefore this device produces velocity dispersion.

The special case $l_w' - l_w'' = 0$ in which the object and image are located at the entrance and exit boundaries, respectively, requires

$$\cos \left[\frac{L}{a_m} \right] = \pm 1 \tag{II-52}$$

when L is $n\pi a_m$ and n is equal to 1, 2, ...

d. Coincident Field Cycloid Mass Spectrometer

The geometry of such an instrument (designed and constructed by Robinson *et al.*), is described in Fig. II-11. It has a focal distance of 6 in. An auxiliary resolving slit with a focal distance of 4.5 in. permits the collection of ions of low mass at an accelerating voltage which is not excessively high. The electric field is produced by distributing potential along a series of 68 window-shaped plates. The cross section of each strip forming the side of the field plate is 100×10^{-3} in². A voltage of 3200 V across the field plate system produces an electric field of the order of 150 V/cm. The magnet pole pieces are welded into the upper and lower plates of the vacuum chamber. A magnetic field of about 10,000 G can be developed in the gap of width of approximately 1.250 in.

The optimum resolution is obtained with injection voltages between 18 and 20% of the field-plate voltage. This gives trajectory excursions between 2.2 and 2.9 in. in the vertical direction and 6.9 and 7.8 in. in the horizontal direction. At low mass the beam width obtained is 0.7×10^{-3} in., which for the S_o of 0.4×10^{-3} in. gives a beam broadening of 0.3×10^{-3} in. due to aberrations. Under typical operating conditions, ion current delivered to the collector per unit

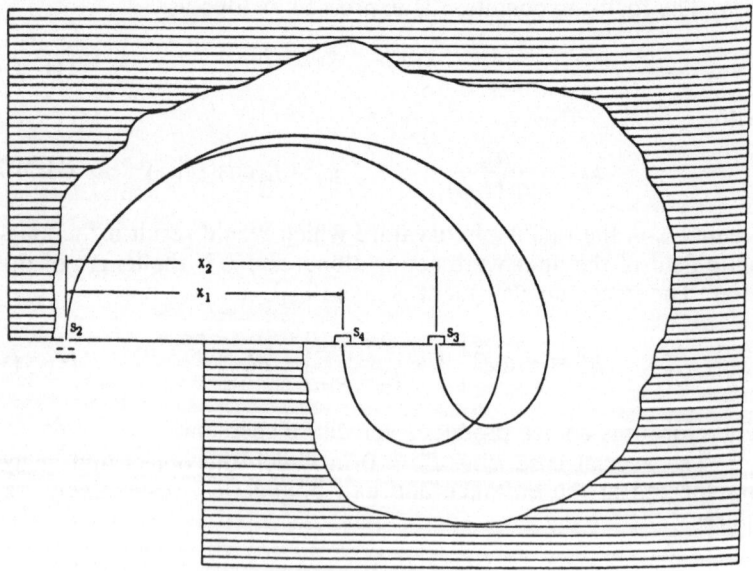

After H. G. Voorhies, *et al.* From *Advances In Mass-Spectrometry*, Pergamon Press, 1959.

Fig. II-11. Geometry of a coincident-field cycloid mass spectrometer.

source pressure per unit ionizing current is 2.3×10^{-12} A/μ/μA. This value is measured for the mass 28 peak of nitrogen at an ion repeller voltage of 130 V, an ion injection voltage of 506 V, and an electron anode current of 40 μA.

e. General Comments Concerning Double-Focusing Instruments

Resolution is the most significant factor in the comparison of double-focusing instruments. Resolution of ordinary double-focusing instruments is approximately equal to S_0/a_e where S_0 is the principal slit width and a_e is the radius of electrostatic analyzer. This may be improved by a factor of two by introducing a Wien velocity filter in the place of a radial electrostatic analyzer.

Resolution, dispersion, and magnification are not mutually independent. For radial electrostatic homogeneous magnetic field combinations,

$$\text{Resolution} \times \text{Dispersion} = \frac{\text{Slit width} \times \text{Magnification}}{100} \quad \text{(II-53)}$$

A compromise is normally achieved between resolution and dispersion. Thus, after the slit width S_0 has been reduced to a tolerable limit, improved resolution is obtained by increasing the dispersion.

The resolution of a given instrument can frequently be improved by reducing the angular speed of the ion beam and the width of the energy bundle entering the magnetic analyzer.

4. DETECTION OF IONS

Fluorescent screens, photographic means, DC amplifiers, vibrating-reed electrometers, electron multipliers, and scintillation methods are some of the means used to detect the resolved ionic masses. We shall discuss briefly the vibrating-reed electrometer and the magnetic electron multiplier.

a. Vibrating-Reed Electrometer

Hull (1932) first drew attention to the usefulness, in the measurement of small DC voltages, of a vibrating condenser acting as an electrostatic generator. Thomas and Finch (1950) and a few others have developed a highly satisfactory detector of positive ions.

The usefulness of the vibrating-reed electrometer is based on the fact that amplification is more easily accomplished with AC than with DC. The input DC potential arising from the passage of the ion current through a large resistor is therefore converted to AC by applying it through a series resistor to a capacitor whose capacitance is periodically varying with DC amplitude. An AC voltage of amplitude $dV = V\,dc/c$, which is proportional to the input DC voltage and which may be fed into a conventional DC amplifier, is thus developed.

In practice, the output of the AC amplifier is generally rectified, using a phase-sensitive device, and applied to the input as negative feedback of sufficient amount to cancel the input voltage. With this null technique, variations in the gain of the amplifier are of secondary importance. The amount of negative feedback indicates the value of the input voltage. Input resistance normally exceeds 10^{15} Ω and input capacitances are approximately 40 pf (picofarads).

b. Magnetic Electron Multipliers for Detection of Positive Ions

As early as 1951, Lincoln G. Smith produced two designs of a 15-stage electron multiplier wherein focusing from one beryllium copper dynode to the next occurs in crossed electric and magnetic fields. This is particularly useful in detecting weak beams or pulses of positive ions in magnetic fields. The first design with dynodes $3/8$ in. wide is usable in fields between about 250 to 460 Oe, whereas the second design with dynodes $1/8$ in. wide is usable in fields between 300 and 1100 Oe. There is a lesser tendency toward ionic feedback, resulting in breakdown and noise, because of uniformity of the electric field. A multiplier could be built with a rise time between 10^{-11} and 10^{-10} sec, which is very probably less than could be obtained with a static multiplier.

In a crossed uniform magnetic and electric field, for an ion of molecular weight M, the transit time over a complete cycle is

$$T = 652 \left(\frac{M}{H}\right) \mu sec \qquad (II\text{-}54)$$

where $T = 2\pi/\omega$ and ω is the cyclotron frequency of the ion. The transit distance is

$$b = 652 \times 10^2 \left(\frac{ME}{H^2}\right) cm \qquad (II\text{-}55)$$

For an electron,

$$T = 0.358 \left(\frac{1}{H}\right) \mu sec \qquad (II\text{-}56)$$

and

$$b = 35.8 \left(\frac{E}{H^2}\right) cm \qquad (II\text{-}57)$$

where E and H characterize the magnitude of electric and magnetic fields.

In any multiplier, it is necessary that the potential of each dynode be of the order of several hundred volts positive with respect to that of the previous one in order that the electrons strike each dynode with sufficient energy to produce further secondaries. This condition is suitably met with proper designing. A uniform electric field is established between one (or two) long "rail" plate (coplanar parallel plates carrying no current) and one (or both) side of the active dynode surfaces which are located in different planes normal to E. Focusing is also very much improved. Figures II-12 and II-13 represent the two types of magnetic multipliers.

In the case of MI, for $H = 450$ Oe, $V = 8110$ V, and $E = 5390$ V/cm, one obtains the ion velocity $\approx 11.98 \times 10^8$ cm/sec and T (for a complete cycle) $= 7.96 \times 10^{-10}$ sec.

In the case of MII, for $H = 1100$ Oe, $V = 8500$ V, and $E = 10,720$ V/cm, one obtains the ion velocity $\approx 9.75 \times 10^8$ cm/sec and T (for one complete cycle) $= 3.25 \times 10^{-10}$ sec.

But there is a finite spread in transit time of the order of 10^{-11} sec. This effect and also the spread in emission time and induction of current in the collector as electrons approach it from the last dynode influence the rise time of a multiplier.

With an unshielded collector, as has been used in both MI and MII, the spread in transit time could contribute to the rise time by an amount of the order of the transit time between dynodes. The use of a plane grid in front of a plane collector not only can maintain uniform E but also make the contribution of the induction effect less than that of the spread in transit time. Thus, a multiplier of high gain and transit time between 10^{-11} and 10^{-10} sec can be real-

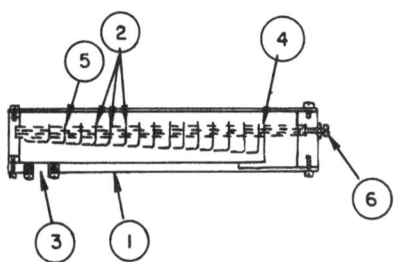

After L. G. Smith, *Rev. Sci. Instr.* **22**, 166, (1951).

Fig. II-12. Magnetic multiplier MI. Central sectional (top) view showing the rail plate (1), dynodes (2), grid through which ions enter (3), collector plate (4), steatite support tubes (5), and compression screws (6).

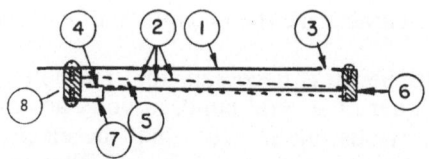

After L. G. Smith, *Rev. Sci. Instr.* **22**, 166, (1951).

Fig. II-13. Magnetic multiplier MII. Central sectional (top) view showing the front and back rail plates (1) and (7), dynodes (2), slit through which ions enter (3), collector plate (4), steatite support tubes (5), and lavite end plates (8).

ized easily. Because of the inherent time focusing accompanying space focusing of electrons executing nearly complete cycles in a magnetic field, the fractional spread in transit time can be considerably reduced.

The magnetic multiplier has tremendous advantages. To obtain a rough comparison of the transit times in the two types of multipliers, we may approximate a stage of the static multiplier by two plane parallel plates between which the same potential difference exists as between successive dynodes in a magnetic multiplier and whose width and distance apart are each equal to b.

The ratio of the transit time (T_m) for an electron starting from rest in the magnetic case to the transit time (T_s) for a similar electron in the idealized static case may be shown to be

$$\frac{T_m}{T_s} = \alpha \left(\frac{\pi y_1}{b}\right)^{1/2} \tag{II-58}$$

where α is the fraction of a cycle completed by the electron in the magnetic case. For multiplier MI, $\alpha = 0.38$, while for MII, $\alpha = 0.87$. Thus, it appears that not only is the fractional spread inherently much greater, but the transit time itself is probably somewhat greater in the static multiplier than in a magnetic one with dynodes of comparable size.

c. A Null Method for the Comparison of Two Ion Currents in a Mass Spectrometer (A. O. Nier, E. P. Ney, *et al.*)

When a mass spectrometer is used to measure the relative abundance of two isotopes, it is customary to set the electric and magnetic fields determining the trajectories of the ions so that ions of one mass will be collected at a particular time. In cases where high resolution is required and the ion currents drift in magnitude or fluctuate widely, a null method for comparing the ion currents is used. Figure II-14 shows the principle of the method symbolically. Positive-ion current i_1 falls on collector 1 and gives rise to a potential drop in i_1R_1 across resistor R_1. A potential V_i appears across the input terminals of the inverse feedback amplifier which has a voltage gain of G. The output voltage is then fed back to the input. Thus,

$$V_i = i_1R_1 - V_o$$
$$V_i = i_1R_1 - G_oV_i$$

Hence,

$$V_i = \frac{i_1R_1}{G_o + 1} \tag{II-59}$$

and

$$V_o = G_oV_i = i_1R_1\left(\frac{G_o}{G_o + 1}\right) \tag{II-60}$$

For very large G_o, $V_o \approx i_1R_1$.

If a positive-ion current i_2 is now allowed to fall on collector 2, which is connected to the grid of a conventional balanced electrometer tube amplifier, a potential drop i_2R_2 will appear across the resistor R_2. If the lower end of the resistor R_2 were grounded, the galvanometer, G, would show a deflection. However, if a bucking voltage of the same magnitude but opposite in direction were applied between B and C, there would be no deflection of the gal-

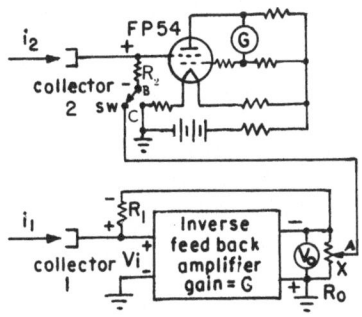

After A. O. Nier, *et al.*, *Rev. Sci. Instr.* **18**, (1947).

Fig. II-14. Instrument for comparing ion currents by the null method.

vanometer. Such a bucking voltage may be obtained by connecting a potentiometer R_o across the output of the feedback amplifier and connecting the tap A to point B. Then the condition for balance of the galvanometer is

$$i_2 R_2 = x V_o = x i_1 R_1 \left(\frac{G_o}{G_o + 1} \right) \qquad \text{(II-61)}$$

or,

$$\frac{i_2}{i_1} = x \left(\frac{R_1}{R_2} \right) \left(\frac{G_o}{G_o + 1} \right) \qquad \text{(II-62)}$$

where x is a fraction of R_o.

The feedback amplifier output thus provides the emf for the potentiometer R_o just as a dry cell would on a conventional potentiometer. But here, the voltage $x V_o$ is automatically proportional to the ion current i_1. Thus the balance point is independent of variations in the ion-current intensities provided both the currents vary in the same way at the same time. In this case, the time constant is $R_1 C / G + 1$ and not $R_1 C$ as it would be in the case of a conventional nonfeedback amplifier.

In this discussion we have tacitly assumed that the ion currents are approximately steady DC. However, if the time constant of the inverse feedback amplifier is adjusted to be the same as that of the FP-54 input circuit, the method can be applied to even sharply pulsating or varying currents. As an example of an application, the comparison of the isotopic abundance ratios in two samples which are very nearly alike can be considered. As measured in an experiment, the abundance ratio of the two isotopes was approximately 100:1 and the corresponding ion currents were approximately 5×10^{-11} and 5×10^{-13} A, respectively. Eight galvanometer readings were made to determine each of the null points and four separate analyses were made of the pair of samples. The differences in the two samples were found to be 0.75, 0.64, 0.82, and 0.76% with a mean deviation of 0.05%.

The null method of measuring ion currents seems to be having extensive applications in addition to making precise determinations of isotope abundance ratios. Relative abundance of two different gases in a complex gas mixture can be compared by this method. By slight variations in the design of the voltage divider R_o, it is possible to read the ratio of the two ion currents automatically and continuously. For this, R_o should form a part of a recording potentiometer. Also, by allowing the recording potentiometer to operate a control device which would hold the ratio of the two gas constituents constant, tremendous industrial applications can be realized.

d. Other Types of Ion Detectors

Flesh and Svec have recently disclosed a model of a mass spectrometer which simultaneously collects positive and negative ions. The ion source consists of two conventional Nier-type sources arranged back to back in which the electron gun is maintained at ground potential and the ion-accelerating electrodes either positive or negative with respect to ground. Positive ions are extracted in one direction and negative ions in the diametrically opposite direction. Mass separation is achieved by means of two 180° magnetic sectors with an ion trajectory radius of 1½ in., and a mass resolution of approximately ⅟₆₀ has been obtained. The mass spectra are scanned magnetically. The model has served well to aid in the solution of circuit problems which will be encountered in a large model of the instrument. Results have been obtained on the mass spectra of chromyl fluoride, CrO_2F_2, Freon 12, CCl_2F_2, etc.

The instrumentation technique adapted by Flesh and Svec is very interesting. Control of the ionizing electron current in the ion source is achieved with an emission regulator circuit. Ionizing

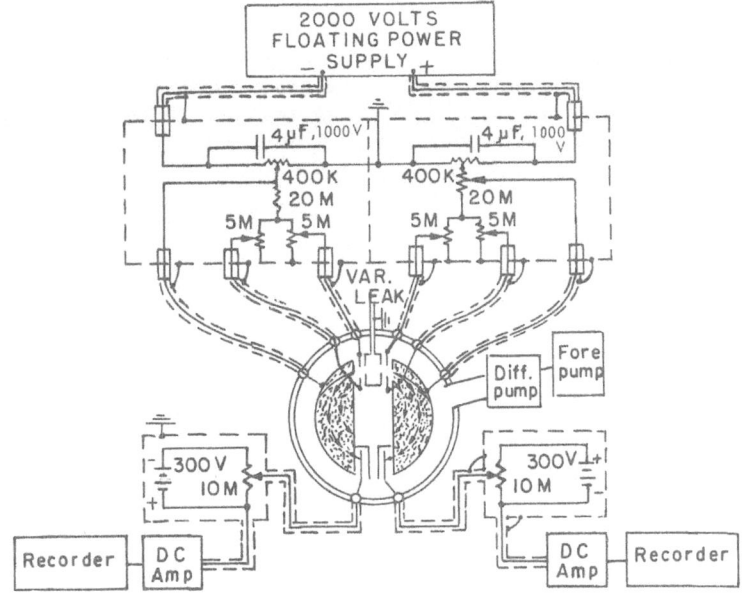

After G. Flesh and H. J. Svec. *Rev. Sci. Instr.* **34**, 399. (1963).

Fig. II-15. Flesh and Svec's ion accelerating and recording circuits.

electron currents in the range 0 to 200 μA at energies from 0 to 150 V are realized.

Figure II-15 represents the schematic diagram of the ion-accelerating and recording circuits used by Flesh and Svec. It appears as though some coupling exists between the two accelerating systems. But it has been experimentally determined that variations from 0 to about 600 V in either the positive or the negative ion section have a negligible effect upon ion currents being collected in the other section. However, capacitive coupling does exist in the form of an apparent ion current whenever the accelerating voltage is varied. Magnetic scanning is achieved by means of a variable current power supply and a scan range of 10 to 200 mass units for 300 V ions is obtained.

A peculiar effect was observed by Flesh and Svec during their measurements. The zero level of the positive-ion amplifier was stable and constant over the entire mass range whereas that of the negative-ion amplifier exhibited a peculiar dependence upon the magnetic field strength. This phenomenon does not seem to be apparent although the currents can be accounted for as due to stray electrons, probably of secondary origin.

However, no attempt has been made to explain the energetics involved in the production of negative ions at the high electron-accelerating voltages. No doubt some of the ions are the result of ion-pair production although in some cases these may be due to decomposition of parent negative ions or even due to secondary, low energy electrons derived from the collision of the primary ionizing electron beam with the walls of the ionization chamber.

Chapter III

Time-of-Flight
Mass Spectrometers

1. INTRODUCTION

The term "time-of-flight" is generally applied to a rather hetero-geneous group of mass spectrometers although, in some instances, the term is a misnomer.

Properly speaking, time-of-flight instruments measure the time required for an ion to traverse a certain specified distance. This may be done either by a direct timing mechanism employing pulsed ion sources and detectors or by subjecting the ions to RF fields. In either case, the apparatus selects from the ion beam ions of a particular velocity. If the velocity of the ion is characteristic of its mass, as in the case of singly charged ions which have fallen through the same potential, such a velocity filter may be used to effect a mass analysis. The first instance in which a velocity filter of this type was employed was in the determination of the velocity of cathode rays by Wiechert (1899).

2. PULSED-BEAM TIME-OF-FLIGHT MASS SPECTROMETERS

a. Theory

Remarkable advances were made during World War II in the development of techniques for the generation of very short elec-trical impulses. These techniques provided the basis for the pulsed-beam mass spectrometers developed independently by Stephens (1946), and Cameron and Eggers (1948).

These instruments consist essentially of a comparatively long drift tube with a source of ions at one end and a detector at the other. The source emits short bursts of ions, homogeneous in either energy or momentum. The ions traverse the drift tube to reach the detector, which is sensitized for a brief instant to register their arrival. Since ions of different masses arrive at the detector at different times, the accurate measurement of the time between activating the source and sensitizing the detector gives information concerning the mass of the ions being detected.

If all the ions have left the source during the accelerating pulse, then all the ions receive essentially the same energy. If the pulse cuts off before any of the ions leave the source, then in a single constant accelerating field, all the ions essentially receive the same momentum.

If L is the length of the drift tube, the transit time for singly charged ions of mass M and constant energy Ve is

$$t = L \left(\frac{M}{2Ve} \right)^{1/2} \qquad \text{(III-1)}$$

The time required for the ions of constant momentum p is

$$t = \frac{LM}{p} \qquad \text{(III-2)}$$

If the collector is sensitized for a period Δt at time t, the resolution for constant-energy and constant-momentum cases is given by:

Constant-Energy Case:

$$\text{Resolution} = \frac{\Delta M}{M} = 2\frac{\Delta t}{t} \qquad \text{(III-3)}$$

Constant-Momentum Case:

$$\text{Resolution} = \frac{\Delta M}{M} = \frac{\Delta t}{t} \qquad \text{(III-4)}$$

It will be noted that the arrangement employing a constant-momentum source possesses, in theory, an advantage of a factor of two over the constant-energy type. However, it is not Δt which in practice limits the resolution but rather the ion thermal energies. When these factors are taken into consideration the advantage may swing in the other direction.

Previous descriptions of the early models of this type of instrument pointed out that they would allow a rapid panoramic display of the entire spectrum and this feature has been emphasized by the use of cathode-ray oscilloscopes as display units. However, the resolutions achieved were poor.

Constant-Energy Case:

$$\frac{\Delta M}{M} = \frac{1}{3} \qquad \text{Cameron and Eggers (1948)}$$

$$= 1 \qquad \text{Keller (1949)}$$

$$= \frac{1}{10} \qquad \text{Wolff and Stephens (1953)}$$

Constant-Momentum Case:

$$\frac{\Delta M}{M} = \frac{1}{20} \qquad \text{Wolff and Stephens (1953)}$$

The time-of-flight mass spectrometer developed by Katzenstein and Friedland (1955) has two important modifications over the previously mentioned constant-energy- and constant-momentum-type instruments. First, a grid placed in front of a continuously sensitive detector is pulsed rather than pulsing the detector itself. The sluggish detector response caused by the capacitance of the collector assembly is therefore eliminated and an experimental resolution of $\frac{1}{75}$ could be achieved. Secondly, provision is made for modulating the electron beam as required by the Fox scheme.

If N is the mass number of the ion, l the length of the drift tube, and U_o and P_o are the energy and momentum of the ion, respectively, then the time of flight t is given by

$$t = l\left(\frac{MN}{2U_o}\right)^{1/2} \qquad \text{(III-5)}$$

for the constant-energy case, and

$$t = l\left(\frac{MN}{P_o}\right) \qquad \text{(III-6)}$$

for the constant-momentum case. Therefore,

$$\frac{N}{\Delta N} = \frac{t}{\Delta t} \qquad \text{(III-7)}$$

in the former case, and

$$\frac{N}{\Delta N} = \frac{1}{2}\frac{t}{\Delta t} \qquad \text{(III-8)}$$

in the latter case.

For constant momentum, an effective energy U_{eff} and thermal velocity v_t are defined as

$$U_{eff} = \frac{P_o^2}{2MN} \qquad \text{(III-9)}$$

and

$$v_t = \left(\frac{2U_t}{MN} \right)^{1/2} \qquad \text{(III-10)}$$

where U_t is the thermal energy of the ions.

The time spread δt in the ion pulses caused by thermal velocities will be

$$\delta t = \pm \left(\frac{U_t}{U_{eff}} \right)^{1/2} t_o \qquad \text{(III-11)}$$

where t_o is normal transit time. Then

$$\frac{\Delta N}{N} = 2 \left(\frac{U_t}{U_{eff}} \right)^{1/2} \qquad \text{(III-12)}$$

and

$$\begin{aligned} t &= \frac{l}{[(2U_o \pm U_t)/MN]^{1/2}} \\ &= \frac{t_o}{(1 \pm U_t/U_o)^{1/2}} \end{aligned} \qquad \text{(III-13)}$$

Applying the binomial expansion

$$t = \tfrac{1}{2} \left(\frac{U_t}{U_o} \right) t_o \qquad U_t < U_o \qquad \text{(III-14)}$$

An instrument to resolve mass number 100 from its neighbors would require a resolution of 0.01. If $l \approx 1\,$m, then a Δt of less than 0.1 μsec will give the desired resolution. Such a time interval is capable of being realized.

The electron beam is accelerated through the ionizing region into a region of high retarding field which brings the beam to rest and returns it through the ionizing region. The use of large parallel plates to mount accelerating grids permits the geometry to be made optimum without the use of narrow electron-defining slits. A schematic representation of the overall instrumentation is shown in Fig. III-1. This figure actually represents the time-of-flight spectrometer developed by Katzenstein. The first six grids of the tube

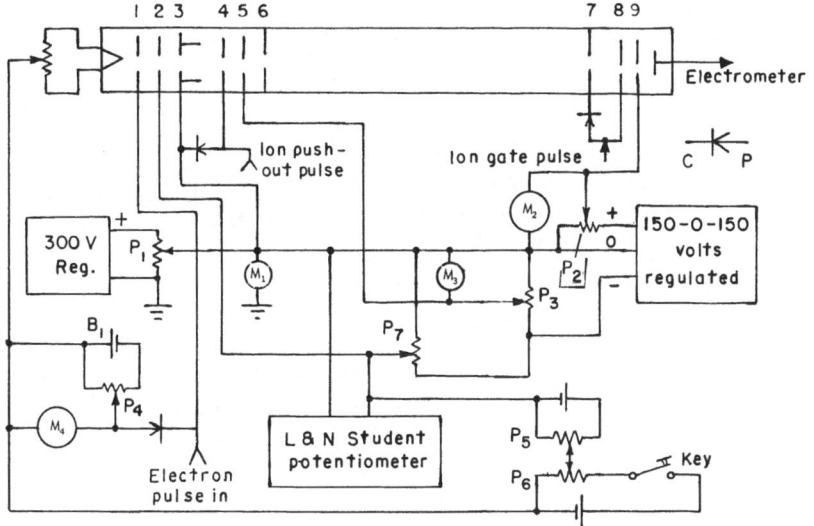

After H. S. Katzenstein and S. S. Friedland, *Rev. Sci. Instr.* **26**, 325, (1955).

Fig. III-1. Schematic representation of the time-of-flight mass spectrometer.

form the ion source. The collector system consists of three grids and a collecting plate. The ion-selector pulse is applied to grid 8 and the retarding potential is applied to grid 9. A block diagram of the timing circuitry used in the mass spectrometer system is shown in Fig. III-2.

The repetition rate oscillator used in the timing circuitry produces 5000 pulses/sec, thereby allowing an ample interval to ensure that all ions due to one pulse are collected before a new pulse is introduced. The pulse from the rate oscillator is sharpened and fed into the electron-pulse generator and to the ion-pulse delay generator. The electron-pulse length and ion-pulse delay are produced by two univibrator gate generators. The output from the delay generator is differentiated, inverted, and used to trigger the ion push-out pulse generator and the ion-selector pulse system. The width of the ion push-out pulse is about 0.2 μsec and is variable in height from 0 to 50 V. The 0.1-μsec ion-selector pulse must be stable, precisely adjustable in time, and free of jitter to the extent of 0.05 μsec.

This instrument appears to be satisfactory as an analytical tool at least up to mass 100 amu (atomic mass units). The ionization

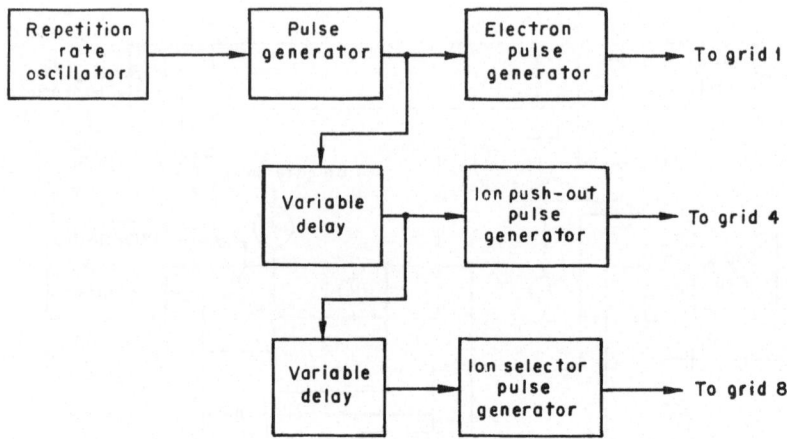

After H. S. Katzenstein and S. S. Friedland, *Rev. Sci. Instr.* **26**, 326, (1955).

Fig. III-2. Block diagram of timing circuitry.

efficiencies obtained for argon and nitrogen using this instrument
are shown in Figs. III-3 and III-4.

b. Wiley – McLaren Time-of-Flight Mass Spectrometer

The Wiley – McLaren time-of-flight mass spectrometer (1955)
has an entirely new type of ion source containing two accelerating

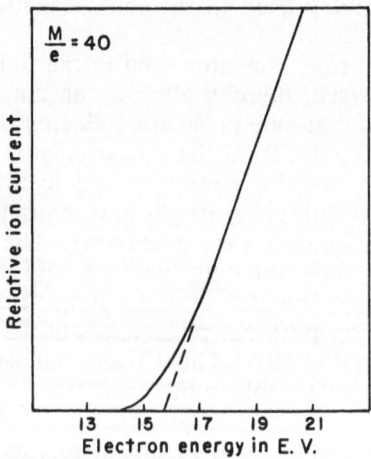

After H. S. Katzenstein and S. S. Friedland, *Rev. Sci. Instr.* **26**, 327, (1955).

Fig. III-3. Ionization efficiency curve for argon.

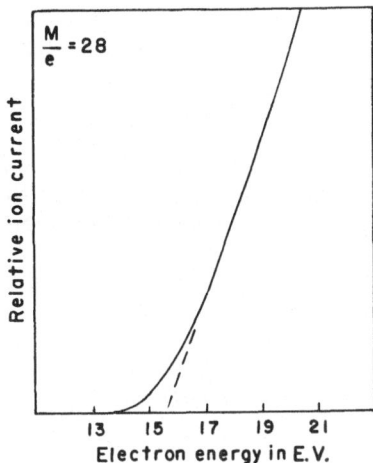

After H. S. Katzenstein and S. S. Friedland, *Rev. Sci. Instr.* **26**, 327, (1955).

Fig. III-4. Ionization efficiency curve for nitrogen.

regions. With the help of this new ion source, a resolution approaching $1/200$ is secured. This mass spectrometer system is shown in Fig. III-5.

While the ions are being formed under pulsed transverse electron bombardment, the potential of the source backing plate is the same as that of the first grid. At all times, there is a field E_d in the second accelerating region d, while the drift space D is field-free. Ions are pushed out of the source toward the collector when a positive pulse giving rise to the field E_S is applied to the source-backing plate. The pulse lasts until all the ions have left the ionization region. This double-field configuration introduces two parameters d and E_d/E_S which are not available in the single-field source.

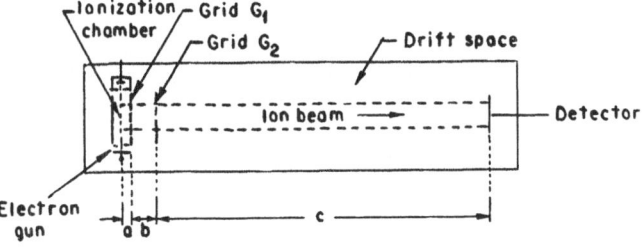

After W. C. Wiley and I. H. McLaren, *Rev. Sci. Instr.* **26**, 1151. (1955) (also courtesy of the Bendix Corporation).

Fig. III-5. Wiley–McLaren time-of-flight mass spectrometer utilizing the new ion source.

Ions are therefore created in different regions of the ionization space with different initial velocities. Disregarding the velocity effect, those ions initially closer to the first grid acquire less energy than, and are eventually overtaken by, those initially closer to the source-backing plate. This constitutes a space focusing or "bunching" which, by adjustment of E_d/E_S, can be made to occur at the collector. An ion initially traveling away from the collector would normally arrive at the collector later than an ion initially traveling toward it.

A Smith type of electron multiplier is used for detection. This possesses a plane conversion dynode which eliminates variations in ion transit time that would be associated with a curved surface.

In moving through the source, any ion with initial energy U_o will increase its energy to a value U.

$$U = U_o + eSE_S + edE_d \qquad \text{(III-15)}$$

Time of flight is given by

$$t(U_o, S) = t_S + t_d + t_D \qquad \text{(III-16)}$$

where

$$t_s = 1.02 \frac{(2m)^{1/2}}{eE_S} [(U_o + eSE_S)^{1/2} \pm U_o^{1/2}]$$

$$t_d = 1.02 \frac{(2m)^{1/2}}{eE_d} [U^{1/2} - (U_o + eSE_S)^{1/2}]$$

and

$$t_D = 1.02 \frac{(2m)^{1/2}}{2U^{1/2}} D$$

For $U_o = 0$ and $S = S_o$, the above equation reduces to

$$t(0, S_o) = 1.02 \left(\frac{m}{2U_t}\right)^{1/2} \left(2k_o^{1/2}S_o + \frac{2k_o^{1/2}}{k_o^{1/2} + 1} d + D\right) \qquad \text{(III-17)}$$

where $U_t = eS_oE_S + edE_d$ and $k_o = (S_oE_S + dE_d)/S_oE_S$.

To find the position at which ions whose initial S values were $S = S_o \pm \frac{1}{2} \delta S$ pass each other, set

$$\left(\frac{dt}{dS}\right)_{0, S_o} = 0$$

and obtain

$$D = 2S_o k_o^{3/2}\left(1 - \frac{1}{k_o + k_o^{1/2}} \frac{d}{S_o}\right) \qquad \text{(III-18)}$$

The focus condition $(dt/dS)_{0,S_o} = 0$ indicates that $t(0,S)$ has either a maximum, minimum, or point of inflection at $S = S_o$. The point of inflection occurs when $(d^2t/dS^2)_{0,S_o} = 0$, which requires, in addition to the focus condition, that

$$\frac{d}{S_o} = \left(\frac{k_o - 3}{k_o}\right) \frac{D}{2S_o} \tag{III-19}$$

If $k_o \leq 3$, then $t(0,S_o)$ is always a minimum point so that for the single-field source $(d = 0, k_o = 1)$ the point $t(0,S_o)$ is a minimum. Utilizing a series expansion of $t(0, S)$ about S_o, we see that

$$\Delta t_{\Delta S} = \sum_{n=1}^{\infty} \frac{1}{n!} \left[\frac{d^n t}{dS^n}(0, S)\right]_{S_o} (\Delta S)^n \tag{III-20}$$

The measure of space resolution M_S is the maximum value of m for which

$$\Delta t_{\Delta S} \leq t_{m+1} - t_m \tag{III-21}$$

where

$$t_{m+1} - t_m = \left[\left(1 + \frac{1}{m}\right)^{1/2} - 1\right] t_m \approx \frac{t_m}{2m}$$

If $k_o \gg 1$ and $k_o \gg d/S_o$, we obtain from equations (III-17), (III-18), and (III-19)

$$M_S \approx 16 k_o \left(\frac{S_o}{\Delta S}\right)^2 \qquad k_o = \frac{D}{S_o} \tag{III-22}$$

Energy resolution is obtained as follows. The time spread Δt_θ introduced by the initial energies is the "turn-around" time of an ion having the maximum initial energy being considered. Now

$$\Delta t_\theta = 1.02 \frac{2 v_o m}{e E_S} \tag{III-23}$$

$$\approx 1.02 \frac{2 (2 m U_o)^{1/2}}{e E_S}$$

and the maximum resolvable mass $M_\theta = t/2\Delta t_\theta$ so that

$$M_\theta = \frac{1}{4} \left(\frac{U_t}{U_o}\right)^{1/2} \left(\frac{k_o + 1}{k_o^{1/2}} - \frac{k_o^{1/2} - 1}{k_o + k_o^{1/2}} \frac{d}{S_o}\right) \tag{III-24}$$

The double-field system has the advantage of requiring a much smaller pulse voltage. The maximum resolvable mass M_θ depends upon both the initial space and energy distributions from

$$\frac{1}{M_{S,\theta}} = \frac{1}{M_S} + \frac{1}{M_\theta} \qquad \text{(III-25)}$$

According to standard specifications used here, $M_S = 400$, $M_\theta = 147$, so that $M_{S,\theta} = 108$ amu. Higher masses may be resolved by using a longer free-flight path and higher energy as well as by the use of time-lag focusing. Large voltages imply good energy resolution, but excessive voltages would complicate the electronic problem of providing the total voltage or the fraction of it that is necessary to produce E_S.

Energy focusing can also be produced by introducing a time lag τ between the formation of ions and application of the accelerating pulse. If v_o is the maximum initial ion velocity, then τ must satisfy

$$\frac{dt}{dS} v_o \tau + \frac{1.02 m v_o}{e E_S} = 0$$

or

$$\tau = \frac{1.02 m}{e E_S (dt/dS)} \qquad \text{(III-26)}$$

The τ and dt/dS combination yields the best overall resolution. The lag systems always give improved energy resolution. In the experimental setup, a magnetic electron multiplier was utilized. The total gain of multipliers used ranged from 10^5 to 5×10^9.

To reduce the background caused by ions which wander out of the ionization region, the electron beam is pulsed on immediately before the ion-accelerating pulse is applied. The beam is controlled by a 100-V pulse lasting for 0.1 to 1 μsec on the electrode nearest the filament. The detected intensity of a single mass group in the spectrometer depends on many spectrometer conditions which are difficult to control precisely (e.g., gas temperature and pressure, ionizing electron-beam intensity, detector efficiency, etc.). However, the measured intensity ratios do indeed remain relatively constant even though wide variations are introduced in the intensity of ions. Figure III-6 represents the experimental computed spectrum of n-butane components using the Wiley–McLaren spectrometer.

An electronic ratio-recording system has also been put into practice by Wiley and McLaren. It contains a multiplier which has, in addition to the final output electrode, two collectors which

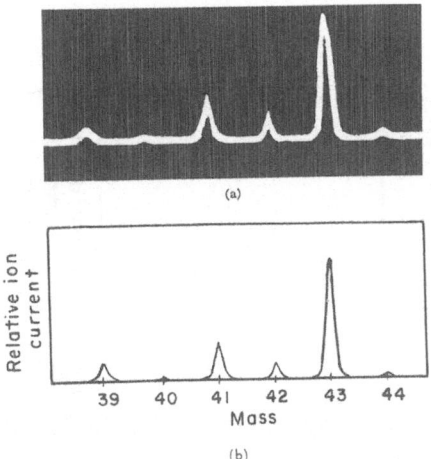

After W. C. Wiley and I. H. McLaren, *Rev. Sci. Instr.* **26**. 1156, (1955) (also courtesy of the Bendix Corporation).

Fig. III-6. Matched experimental and computed spectrum (n-butane components); from left to right, masses 39, 40, 41, 42, 43, and 44 amu. The peak width is 0.024 μsec. (a) Experimental mass spectrum. (b) Computed mass spectrum.

can be activated electronically. A mass peak is accepted by one of these collectors if a short-duration pulse of 60 V is applied to its "gating" electrode. The multiplier collector electrodes are each connected to integrating DC amplifiers. The amplifier outputs are connected to a recorder which directly records the ratio.

c. Ion Velocitron

The ion velocitron developed by Cameron and Eggers (1948) belongs to the pulsed-beam type of instrument and the time interval between the arrival of pulses of ions of masses m_1 and m_2 at the end of the tube is proportional to $L[(m_1)^{1/2} - (m_2)^{1/2}]$ where L is the length of the drift tube from ion source to collector plate. The resolution between ions of mass difference $\Delta m = m_1 - m_2$ is determined by the accelerating voltage, length of drift tube, and by the width in microseconds of the impulse which can be produced initially and resolved by the pulse amplifier. The source must produce ions which are nearly monoenergetic, since broadening of the received pulse will occur if this condition is not fulfilled.

The experimental setup is illustrated in Fig. III-7. The last deflecting plate is at ground potential and the first two could be varied independently. The first plate is normally operated at 300 V

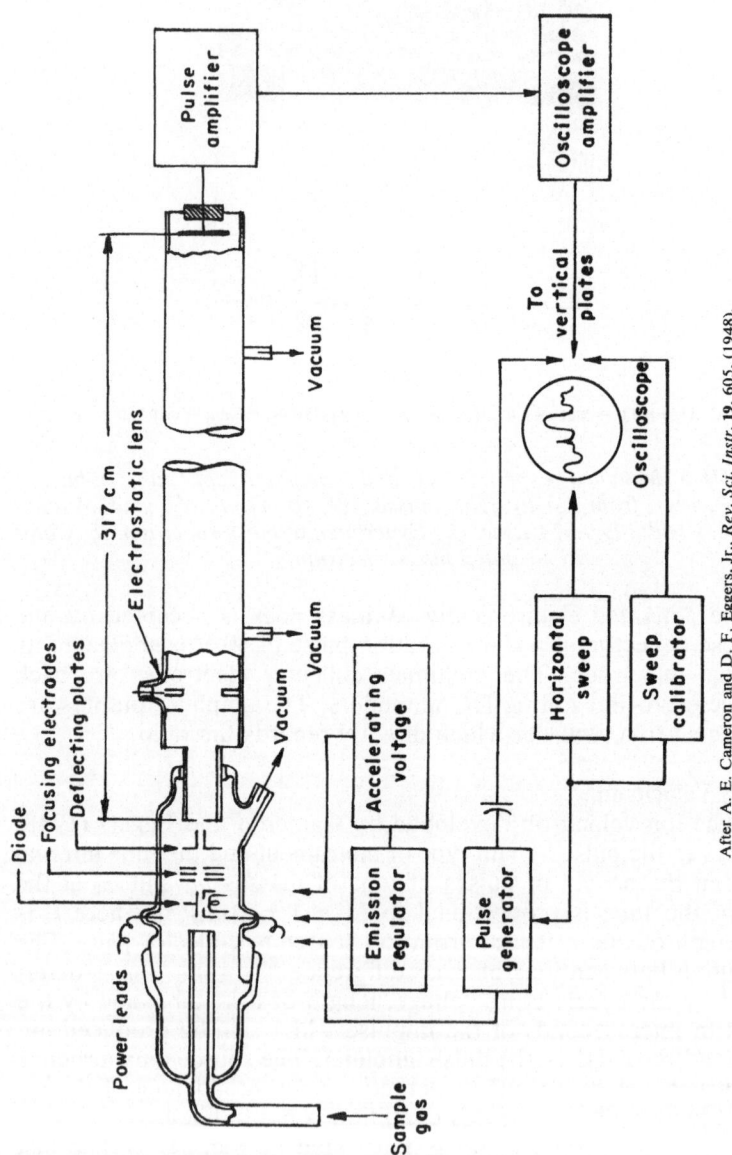

After A. E. Cameron and D. F. Eggers, Jr., *Rev. Sci. Instr.* **19**, 605, (1948).

Fig. III-7. Cameron and Eggers' ion velocitron.

positive and the second plate at about 500 V negative with respect to ground. The four-stage pulse amplifier has a high-frequency pass of 1 Mc and an input resistance of 1.6 MΩ in order to reduce the time constant of input circuit and thereby increase the sensitivity. This amplifier is coupled to the transmission line through a cathode-follower stage. The patterns obtained on the screen could be photographed using very sensitive film (like orthochromatic or panchromatic types).

The patterns are analyzed by reading the time corresponding to the midpoint of rise of the peak. Masses are calculated from the observed time intervals by the expression

$$m = \frac{t^2 E}{5.25 \times 10^4} \tag{III-27}$$

where E is in volts and t is in microseconds. Sweep calibrator pulses could be injected at regular time intervals and this shows up conspicuously as marker pips at regular time intervals on the horizontal trace of the mass-spectrum display. Figure III-8 shows the oscilloscopic pattern obtained with mercury vapor. The last one shows the marker pips spaced 10 μsec apart.

3. RADIO-FREQUENCY TIME-OF-FLIGHT MASS SPECTROMETERS

In contrast to the low transmission of the pulsed-beam machines, the radio-frequency TOF (time-of-flight) machines, whether

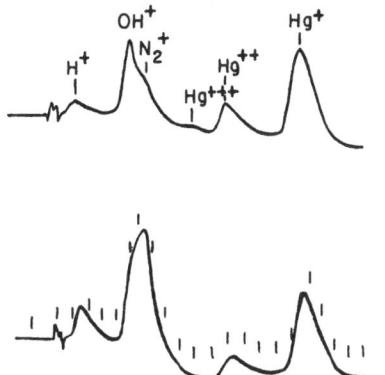

After A. E. Cameron and D. F. Eggers, Jr., *Rev. Sci. Instr.* 19, 605, (1948).

Fig. III-8. Hand-drawn oscilloscopic pattern of mercury vapor with marker pips.

they are focused or not, have very high transmissions of the order of 10 to 50%. However, the Bennett RF mass spectrometer has very low transmission even with the poor resolution obtained.

a. Hipple, Sommer, and Thomas Mass Spectrometer

A focusing-type RF TOF mass spectrometer developed by Hipple, Sommer, and Thomas is illustrated in Fig. III-9. This instrument is actually a small cyclotron adapted to sharpen mass discrimination. The ionizing electron beam is directed along the axis of the tube parallel to the magnetic field. If one places an RF field at right angles to the magnetic field, the ions are accelerated by that field and follow roughly circular trajectories. If the period of rotation is equal to the RF period, ions will be successively accelerated and spiral outward to the collector. The resonant frequency is a measure of the mass of ions being collected.

The resolving power of such an analyzer is proportional to the number of revolutions the ion makes before reaching the collector. Loss in resolving power is due largely to space charge built up in opposition to the trapping field, to inhomogeneities in the magnetic field, and to gas scattering effects. To minimize these effects, it is important that the machine be small. To make the machine small and still keep the number of revolutions high, it is necessary to use very intense magnetic fields and very small RF voltages.

For practical purposes the machine is limited to rather low mass work. It is an extremely sensitive instrument with high resolving power for masses of about 30 and below. Above these masses, it cannot very well compete with static field machines.

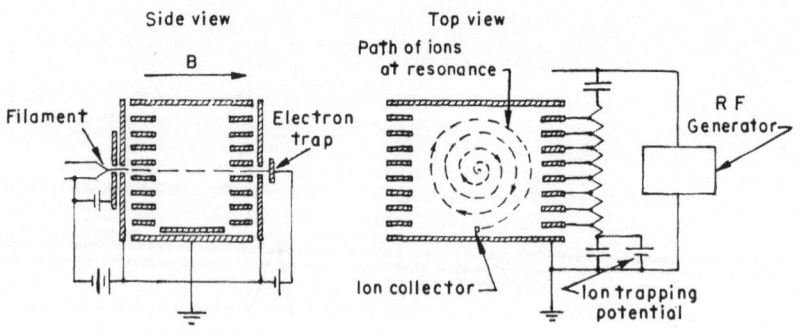

After J. A. Hipple, *et al.*, *Phys. Rev.* **76**, 1877, (1949); **78**, 332, (1950).

Fig. III-9. Hipple, Sommer, and Thomas' time-of-flight mass spectrometer.

b. Glenn's Mass Spectrometer

The nonfocusing RF TOF mass spectrometer developed by Glenn (1952) is shown in Fig. III-10. Figure III-11 represents the circuit interconnections. The ions formed at S are accelerated to B and the acceleration voltage is such that the transit time between the bunching grids B and B' is short in comparison to the period of voltage change on the bunching grids.

The accelerating voltage required to satisfy the experimental conditions can be theoretically determined. Let t be the time at which ions leave the buncher grids, τ be the time of arrival at the gate grids G' and G, and L be the distance between the buncher grids and gate grids. The theory involves the determination of the accelerating voltage $V(t)$ such that ions leaving the buncher grid at time t arrive at the gate grid at time τ independent of t. The velocity of ions is given by the equation

$$v = \frac{L}{\tau - t} = \left(\frac{2eV}{m} \right)^{1/2}$$

$$\approx 1.396 \times 10^6 \left(\frac{V}{m} \right)^{1/2}$$

(III-28)

where $(\tau - t)$ represents the transit time.

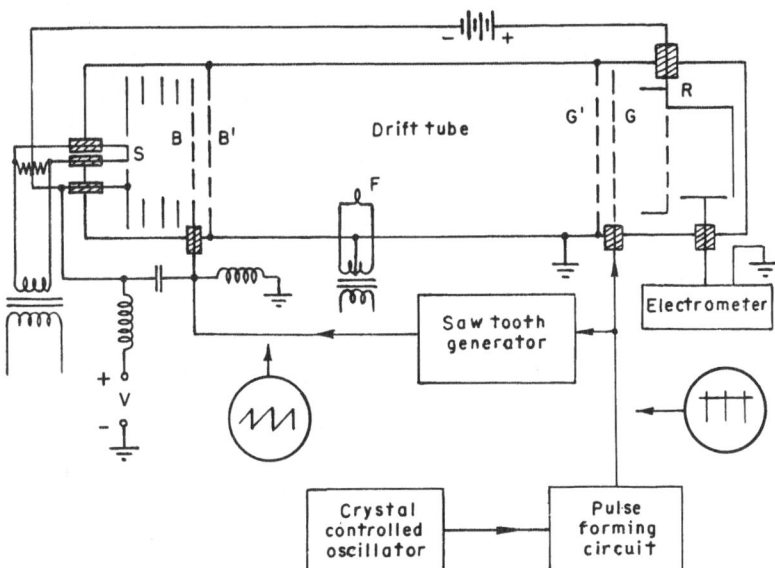

After W. E. Glenn, U.S. AEC Report, AECD-3337-1952, and also UCRL-1628. Courtesy of U.S. Atomic Energy Commission and also of the University of California, Lawrence Radiation Laboratory, Berkeley, California.

Fig. III-10. Glenn's time-of-flight mass spectrometer.

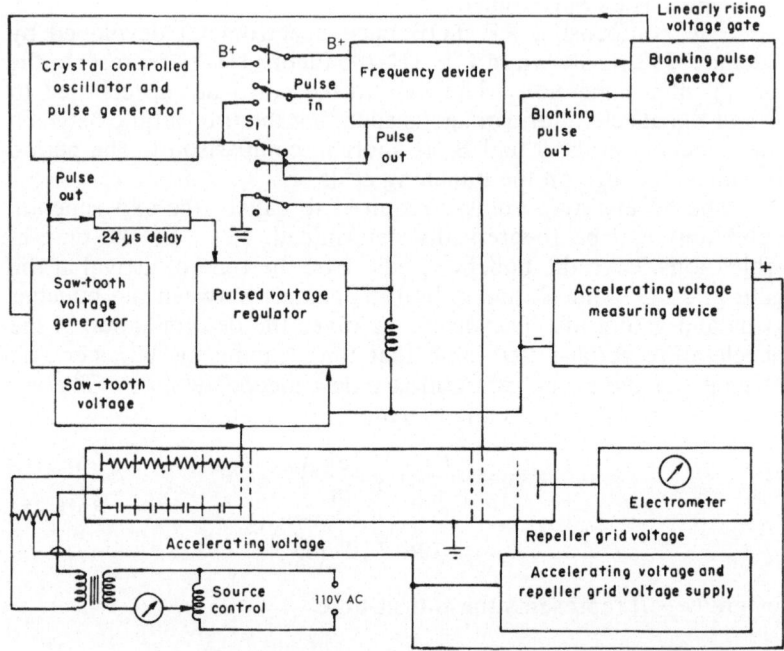

After W. E. Glenn, U.S. AEC Report, AECD-3337-1952, and also UCRL-1628. Courtesy of U.S. AEC and
also of the University of California, Lawrence Radiation Laboratory, Berkeley.

Fig. III-11. Block diagram of circuit interconnections.

Solving for V, we obtain

$$V(t) = \frac{mL^2}{2e\tau^2} \sum_{\eta=0}^{\infty} (\eta + 1)\left(\frac{t}{\tau}\right)^{\eta} \qquad \text{(III-29)}$$

In order that ions which leave the buncher grids in the interval
$0 \leqslant t \leqslant t_1$ all arrive at the gate grids at the time τ, it is necessary
that V varies with time according to the above equation. If $t_1 \ll \tau$,
then

$$V(t) \approx \frac{mL^2}{2e\tau^2}\left(1 + \frac{2t}{\tau}\right)$$

$$\approx 5.11 \times 10^{-13} m \left(\frac{L}{\tau}\right)^2 \left(1 + \frac{2t}{\tau}\right) \qquad \text{(III-30)}$$

The form of voltage to be applied between B and B' must therefore be linearly increasing. In order to maintain the linearly rising form of the voltage term, each bunch is cut off after a short time and a new bunch is started, i.e., a saw-tooth RF bunching voltage is used. Analysis of the bunched beam is accomplished by pulsing the gating grid at time τ to permit that bunch to pass through to the detector and to repulse groups of different mass. The inevitable harmonic peaks have to be overcome at the cost of transmission. The simplest form of harmonic suppressor amounts to changing the machine to a pulsed-beam machine. In some applications, it may be desirable to hold the accelerating voltage constant and vary the transit time. In this case, the transit time is proportional to the square root of the mass collected.

In discussing the errors in mass determination, Glenn lists a series of errors while analyzing his instrument:

$$\text{Bunching error} \left(\text{with } \frac{t'}{\tau} = \frac{1}{40} \right) \cdot \cdot \frac{1}{2130}$$

$$\text{Thermal error} \ldots \ldots \ldots \frac{1}{563}$$

$$\text{Saw-tooth voltage error} \ldots \ldots \frac{1}{2000}$$

$$\text{Gate pulse error} \ldots \ldots \ldots \frac{1}{900}$$

$$\text{Length error} \ldots \ldots \ldots \ldots \frac{1}{865}$$

These errors are evaluated with reference to a mass of 250 mass units and are in terms of mass units. The summation of errors will then be approximately

$$\Delta = \left[\left(\frac{1}{2130}\right)^2 + \left(\frac{1}{563}\right)^2 + \left(\frac{1}{2000}\right)^2 + \left(\frac{1}{900}\right)^2 + \left(\frac{1}{865}\right)^2 \right]^{1/2}$$

$$= \frac{1}{405} \tag{III-31}$$

$$= 0.62 \text{ mass units}$$

which is sufficiently small. For masses greater than 107, Glenn also observed that the accelerating voltage must be above approximately 725 V for a negligible loss in transmission due to thermal velocities alone. For lower masses, a higher accelerating voltage is needed to obtain high transmission but this will then reduce the percentage resolution.

Harmonic elimination is one of the major problems in the design and operation of time-of-flight instruments. Some of the systems designed for that purpose had to be rejected because of their inherent inability to overcome the DC background current caused by the harmonics which may vary more than the current of low-abundance mass peaks. Another group of harmonic-elimination systems had to be rejected because of the complexity of design and construction, necessity of maintaining high magnetic fields, and reduction of transmission due to fringing fields.

Two of these rejected instruments are (1) a low-resolution magnetic mass spectrograph in cascade with the time-of-flight instrument, and (2) crossed electrostatic and magnetic fields used as a velocity selector that would leave undeflected only ions of a velocity corresponding to the desired transit time.

Glenn developed a successful harmonic-eliminator system which automatically reduces the transmission of the spectrometer instrument. The main purpose of the new system would be to get rid of the unwanted ion bunches which would otherwise give harmonic peaks. For this purpose, a square pulse is added to the linearly rising bunching voltage. This can be called the *blanking pulse*. The wave form is illustrated in Fig. III-12. A DC voltage equal to the height of the square pulse is also added to the repeller grid (R) voltage. Only ions bunched during the pulse have enough energy to overcome this repelling voltage. Since there is only one gate pulse per transit time, the only ions that are collected are those of the proper transit time that were bunched during the blanking pulse. The voltage is low across the bunching grids during the bunching of the ions that will be selected so that transverse velocities due to fringing fields will be reduced. This is decidedly advantageous.

Glenn reported that the transmission obtained was 20% with the harmonic eliminator off and this value was reduced to 0.5% with the harmonic eliminator on. Theoretically and experimentally, this type of mass spectrograph can match the conventional mass spectrograph in cases where transmission, stability, and reproducibility are important. Also, this instrument is less expensive and highly portable compared to the conventional mass spectrographs. Figure III-13 represents the current versus mass curves for rubidium and cesium taken with Glenn's mass spectrograph.

c. Smythe–Mattauch Time-of-Flight Mass Spectrometer

Smythe (1926) built this mass spectrometer at the California Institute of Technology. In this scheme, alternating electric fields

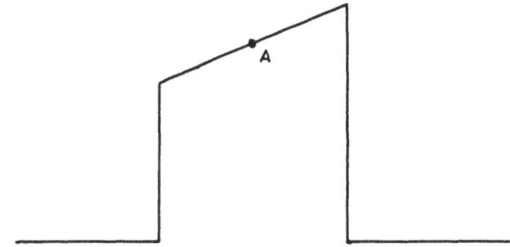

After W. E. Glenn, U.S. AEC Report, AECD-3337-1952, and also UCRL-1628. Courtesy of U.S. AEC and also of the University of California, Lawrence Radiation Laboratory, Berkeley.

Fig. III-12. Wave form of blanking pulse.

are applied at right angles to an ion beam in such a way that only those particles having certain velocities emerge from the fields undisplaced and undeflected.

Figure III-14 represents the schematic diagram of the electrode system in the spectrometer. An alternating voltage of frequency f is applied to the plates of the two condensers A and B, each of length $2a$. If the resulting deflecting fields stop abruptly at the physical boundaries of the plates, then those ions with velocity v_0 such that they pass through condenser A in an integral number of cycles n_S receive as much acceleration in one direction as in the other, and upon emerging, follow paths which are parallel to their original direction of motion, albeit displaced. In the second identical condenser B which the ions enter in the opposite phase to that in which they entered the first, the displacement is reversed and the particles emerge from the combination both undeviated and undisplaced. However, in practice, the deflecting fields are neither uniform nor sharp-edged.

For simplicity, Smythe and Mattauch made $f(x) = f(-x)$. This was done by forming both A and B from pairs of identical condensers, placed in identical chambers. This simplification leads to the transmission of an unanticipated number of velocities corresponding to various n's. These "ghosts" which in the earlier stages of the development of the instrument limited its usefulness have since been completely explained by Hintenberger and Mattauch (1937).

The beam emerging from the filter is further analyzed by a radial electrostatic field and the resolution achieved is quite adequate to permit a useful study to be made of the abundance of various isotopes. Smythe and Mattauch could study the abundance of O^{18}.

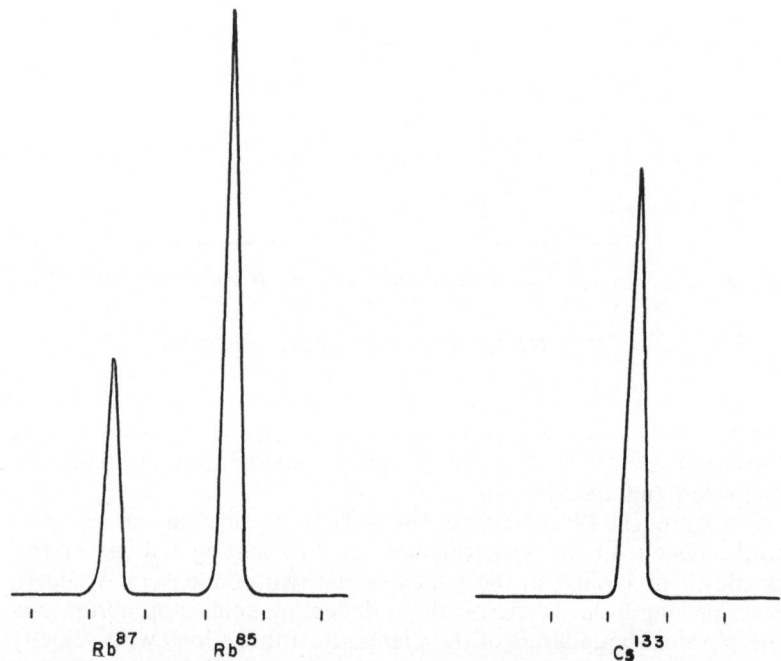

Rb^{87} Rb^{85} Cs^{133}

After W. E. Glenn, U.S. AEC Report, AECD-3337-1952, and also UCRL-1628. Courtesy of U.S. AEC and also of the University of California, Lawrence Radiation Laboratory, Berkeley.

Fig. III-13. Current vs. mass curves of rubidium and cesium.

4. A NEW MAGNETIC ELECTRON MULTIPLIER FOR DETECTION OF IONS IN A TIME-OF-FLIGHT INSTRUMENT

The ion bunches received at the detector of a time-of-flight mass spectrometer may have a very short duration, so wide-band amplification techniques are necessary. A magnetic electron multiplier designed by Goodrich and Wiley satisfies this purpose. It uses a strip of semiconducting material for the multiplying surface instead of dynodes. A gain of 10^7 is obtained with a dark current of only 3×10^{-21} A. The output signal can be displayed on a cathode-ray oscilloscope, the time base of which is triggered by the spectrometer pulse generator. Thus, the mass spectrum can be obtained photographically. The mass spectrum is also recorded on an automatic recorder with suitable input connections from the detector. By the use of a gating system the electron beam from the multiplier is directed toward an anode during the period of each cycle when one particular mass is being received at the detector. The anode signal is fed to the recorder. If the delay of the gate pulse is varied continuously, the recorder traces out a mass spectrum. In order to eliminate drifts in peak heights, two anodes are gated, one with a fixed delay and one with a variable delay. The first gate pulse is set on a reference mass, and a feedback system controls the gain of the multiplier so that the anode signal is maintained at a constant level. A stabilized mass spectrum is obtained from the second anode by varying the delay of its gate pulse.

5. ADVANTAGES AND DISADVANTAGES OF TIME-OF-FLIGHT MASS SPECTROMETERS

The main advantages of TOF spectrometers are (1) the speed with which a spectrum can be obtained conveniently, (2) the ability to record entire mass spectrum for each accelerating pulse, (3) their dependence for accuracy on electronic circuits rather than on extremely critical mechanical alignment and on the production of highly uniform, stable magnetic fields, and (4) the fact that a TOF spectrometer with an oscilloscope display records all the mass peaks which are originally formed by the source, and loses only those stray ions which emerge from the source at an angle which causes them to miss the ion detector. A conventional spectrometer detects only one peak at a time.

The main disadvantage of nonmagnetic TOF spectrometers has been their limited resolution. In practice, the resolving power of a TOF spectrometer depends on its ability to reduce the time spread caused by the ever present initial space and initial kinetic energy distributions. The ability of the spectrometer to resolve masses despite the initial space distribution is called *space resolution*, while its reduction of the time spread introduced by the initial kinetic energy distribution is called *energy resolution*.

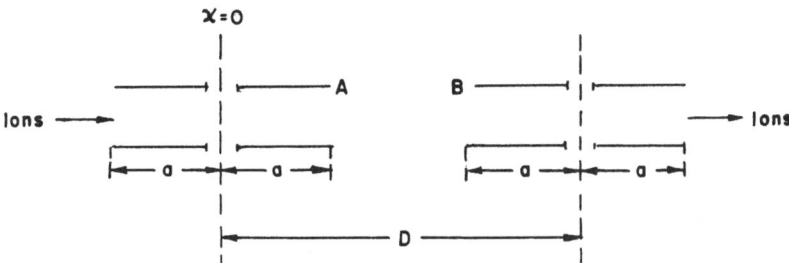

From *Mass Spectroscopy* (H. E. Duckworth), Cambridge University Press, 1958.

Fig. III-14. Schematic diagram of electrode system in the Smythe–Mattauch time-of-flight mass spectrometer (RF type).

The space resolution problem is to reduce the time spread $\Delta T_{\Delta S}$ where ΔS is the space deviation. The maximum time spread introduced by initial velocities, ΔT_θ, is the difference in flight times between a pair of identical ions formed at the same S position with the same maximum initial spread but with oppositely directed velocities. ΔT_θ can be minimized by remodeling the electron gun. A compromise between space and energy resolutions is struck in designing an ion gun to obtain improved resolution.

6. COMPARISON BETWEEN THE TIME-OF-FLIGHT MASS SPECTROMETERS AND THE MAGNETIC-TYPE INSTRUMENTS

Stability and transmission are the two fields in which wide comparison can be struck between the TOF and the magnetic instruments. In reference to stability, the TOF instrument has more flexibility and accuracy due to the replacement of conventional magnets by electronic oscillators. In a conventional type, the magnetic field is hard to regulate and measure. Also, it is very difficult to calibrate a magnetic instrument due to fringing fields, hysteresis, and saturation effects of the core. A TOF instrument is extremely sensitive. In either instrument, a direct proportionality exists between the accelerating voltage and mass if the magnetic field or frequency is held constant.

When comparing transmission, though this factor is low in both types of instruments, TOF instruments definitely have a higher transmission factor compared to conventional types (perhaps on the order of 25 times greater).

Other comparisons may be made. The magnetic instrument lends itself more readily to the collection of several masses simultaneously. It is very difficult to obtain the same result with a TOF instrument. The TOF instrument has so-called "harmonics," not found at all in the magnetic instruments. Elimination of harmonics with still reasonable transmission is very difficult with a TOF instrument. The upper limit of beam current is about the same for the 60° magnetic instrument and the TOF instrument. The 180° magnetic mass spectrograph can handle considerably higher currents since the entire trajectory is in the magnetic field and a plasma of electrons is set up along the ion path which tends to neutralize space-charge effects.

Chapter IV

Radio-Frequency
Mass Spectrometers

1. INTRODUCTION

In these instruments, certain ions with a particular initial velocity are accelerated by radio-frequency fields to the extent that they are able to overcome a DC potential barrier and reach the collector. Other ions experience insufficient acceleration in the RF field to surmount the final retarding voltage and hence are stopped.

In an instrument suggested and first constructed by Smythe (1932), the defocusing of the unwanted ions is achieved progressively as the ions travel toward the collector.

As shown in Fig. IV-1, the ion beam passes between the plates of two parallel plate condensers supplied with RF. Each condenser is actually divided in two as shown in the figure to produce a field shaping which seemed desirable on the basis of Smythe's calculations. It was shown that for a given RF, the transverse fields would produce no lateral displacement of ions having a certain velocity and these ions could therefore pass through the second slit and reach the collector. In practice, however, it was found that instead of a single velocity band, there were several separate ion velocity bands which would permit ions to reach the collector, thereby giving rise to "ghost" peaks in the mass spectrum. The design seems to have been abandoned for this reason.

By far the most common type of nonmagnetic mass spectrometer produces defocusing of the unwanted ions by means of a DC electrostatic lens placed in front of the collector. Since the lens is sensitive only to the energy of ions entering it, and not to the e/m ratio, the ions in the beam must have an energy distribution

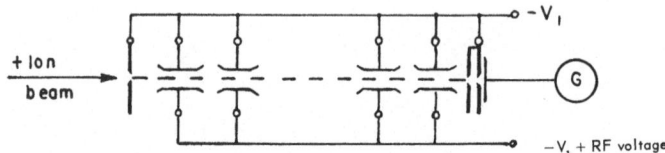

After L. W. Kerr, *J. Electronics* **2**, 182, (1956). Courtesy of Taylor and Francis, Ltd.

Fig. IV-1. Mass spectrometer using the principle of progressive defocusing (Kerr).

which depends on their e/m ratio. An electrode structure giving the ions this property is called an *analyzer* and mass spectrometers using this defocusing principle are called *energy gain instruments*.

The defocusing lens generally used is of a very simple type, consisting of a grid placed in front of the collector, with a retarding potential V_R applied relative to the ion source. Only those ions which gain at least V_R volts in passing through the analyzer system can therefore reach the collector, and all other ions are heavily defocused. The simplest analyzer is the δ-function type (Boyd, 1950) which is shown with a defocusing lens in Fig. IV-2.

In the simplest case, no RF signal is applied to grid system A and a small RF voltage $V \cos(2\pi t/\tau)$ is applied to grid system B. It is clear that the ions passing through the first pair of grids (or "gap") at time $t = 0$ with a velocity $v = 2L/\tau$ will pass through every gap at the moment of peak accelerating field and hence gain NV volts in traversing an N-gap analyzer. Ions of differing velocities cannot, in general, pass through every gap at the optimum RF phase angle and so must gain less energy than NV. Those ions with velocities equal to $2L/\tau$ are called "tuned" ions, and by using a value of $V_R \approx NV$, only those ions whose velocities lie close to that of tuned ions are allowed to reach the collector.

Some difficulty arises however from a special class of untuned ions called harmonically tuned ions, whose velocities are such that they have a transit time of $3\tau/2$, $5\tau/2$, $7\tau/2$, etc., between gaps. Such ions can also pass through every gap at the moment of peak field and will reach the collector together with the tuned ions. Hence a perfect δ-function analyzer cannot separate singly charged ions of mass M from those of mass $9\,M$, $25\,M$, $49\,M$, etc.

In practice, analyzers of the δ-function type are imperfect in that the gaps cannot be made infinitesimally small so that ions do not pass instantaneously through a gap, but take a finite time during which the RF field changes. This rise and fall of the field is more important for the slowly moving harmonically tuned ions, so that

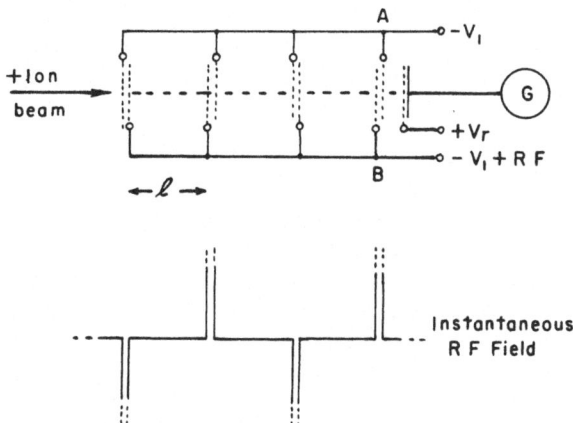

After L. W. Kerr, *J. Electronics* **2**, 183, (1956). Courtesy of Taylor and Francis, Ltd.

Fig. IV-2. Energy gain mass spectrometer of the δ-function type (Kerr).

while tuned ions can gain almost NV volts, the maximum energy gain of harmonically tuned ions is rather less, and a sufficiently large value of V_R will prevent them from reaching the collector. Therefore, a drastic reduction in current efficiency results since V_R must lie so close to NV that only those tuned ions which happen to enter the first gap when the RF field is close to the peak value can gain an energy greater than V_R and hence reach the collector.

2. BENNETT-TYPE RADIO-FREQUENCY MASS SPECTROMETER

In 1950 Bennett designed an RF mass spectrometer based on the velocity selection method. The tube has parallel plane grids constructed of knitted wire nets with a large percentage of open area and arranged in groups of three. An RF alternating potential is applied to the middle grid of each group and stopping potentials are used to reject all ions except those with the selected mass. The spectrometer possesses sufficient resolution for ordinary gas analysis requirements, and is simpler, more compact, and more rugged than magnetic-beam deflection devices. Bennett's new approach grew out of some observations made in 1946 on a pentode with coaxial cylindrical electrodes. An RF potential was applied to the intermediate one of the three grids while negative ions were produced at the hot cathode in the central portion of the tube. Although this tube was built for the purpose of separating negative ions differing in mass by a factor of two or more, it was then realized that a much better separation had been effected than had been anticipated.

a. Single-Stage Tube

Figure IV-3 is a representation of a single-stage tube of Bennett. Electrons from the filament a are accelerated through a DC potential to the first grid b. Positive ions produced by collisions between these electrons and gas molecules in the tube between grids b and c are accelerated towards c due to the potential difference V between b and c. The grids c, d, and e are equally spaced at a distance S and an RF field of angular frequency ω is applied to the middle grid d, so that the electric field between grids c and d is given by

$$E_{cd} = -E \sin(\omega t + \theta) \qquad \text{(IV-1)}$$

and the electric field between grids d and e is given by

$$E_{de} = -E \sin(\omega t + \theta) \qquad \text{(IV-2)}$$

where $t = 0$ at the instant when an ion crosses the plane of grid c, and $E \sin \theta$ is the value of the field E_{cd} at that instant. If one assumes a uniform flow of ions to grid c, then there will be an equal number of ions for each possible value of θ. Some of these ions will

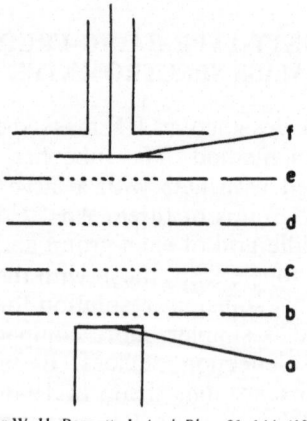

After W. H. Bennett, *J. Appl. Phys.* **21**, 144, (1950).

Fig. IV-3. Single-stage tube (Bennett).

receive energy from the fields as they pass the grids while others will lose energy to the fields. The collecting plate f has a DC stopping potential which rejects all ions except those which have acquired nearly the maximum possible energy from the fields and which originated near the grid b.

It is assumed that the change in velocity of an ion as it moves through the grids is small compared to its initial velocity so that the effects of the applied fields on the transit time may be neglected.

The force on a mass m between grids c and d is

$$F_{cd} = m\ddot{x} = eE \sin (\omega t + \theta) \tag{IV-3}$$

and the force between grids d and e is

$$F_{de} = m\ddot{x} = - eE \sin (\omega t + \theta) \tag{IV-4}$$

where t is measured from the time the ion passes the grid c and θ is the phase angle of the alternating potential on d at the instant the ion passes c. Therefore the energy acquired by the ion from the alternating field is

$$\Delta W = \Delta (\tfrac{1}{2}mv^2) = v \cdot \Delta (mv) = v \int F\, dt \tag{IV-5}$$

The transit time for the ion to travel from one grid to the next is S/v where S is the distance between grids.

Then,

$$\Delta W = v \left[\int_0^{S/v} eE \sin (\omega t + \theta)\, dt + \int_{s/v}^{2S/v} - eE \sin (\omega t + \theta)\, dt \right]$$

so that

$$\Delta W = \frac{eEv}{\omega} \left[\cos \theta - 2 \cos \left(\frac{S\omega}{v} + \theta \right) + \cos \left(\frac{2S\omega}{v} + \theta \right) \right] \tag{IV-6}$$

ΔW is a maximum with respect to variations in θ when

$$\frac{S\omega}{v} + \theta = \pi = 180° \tag{IV-7}$$

This shows that ions which pass grid d just as the field reverses obtain maximum incremental energy. ΔW is a maximum with respect to variations in ω when $\theta = 46°$ when the transit angle between the grids is $S\omega/v = 134°$. The velocity of the ion receiving maximum incremental energy is obtained from

$$eV = \tfrac{1}{2} M m_0 v^2 \qquad\qquad \text{(IV-8)}$$

where V is the potential difference between the ion source b and the grid (c), M is the mass number of the ion in atomic mass units, and m_0 is the mass of an ion of unit mass number. After substituting and rearranging

$$M = \frac{0.266 \times 10^{12} V}{S^2 f^2} \qquad\qquad \text{(IV-9)}$$

Figure IV-4 shows the energy ΔW acquired by an ion passing the grid d at 180° phase as a function of the number of cycles N executed by the alternating field while the ion travels from grid c to grid e. The principal maximum of this curve is at $N = 0.74$ cycle because the transit angle from the first to the third grid for maximum ΔW is

$$\frac{2S\omega}{v} = 267° = 0.74 \text{ cycle} \qquad\qquad \text{(IV-10)}$$

If the stopping potential applied in the collecting system corresponds to the energy Z, the ions can be collected only at frequencies corresponding to positions in the figure between A and B. Ions passing the first grid at phase angles different from the optimum 46° angle acquire a lower energy and can reach the collecting electrode within a narrower range than the range between A and B. The ions collected are those at all phases of entry which can collect enough ΔW to exceed Z. So, the shape of the line observed for ions of a particular mass is sharper than that portion of the curve above

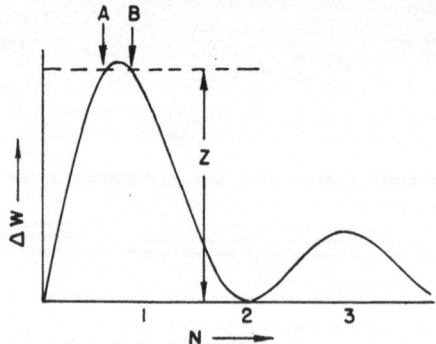

After W. H. Bennett, *J. Appl. Phys.* **21**, 145. (1950).

Fig. IV-4. ΔW vs. N, single stage (Bennett).

Z. In Fig. IV-5, graphically depicting ion currents for positive ions of mercury, the upper curve refers to the case when a smaller stopping potential is applied as compared to that in the case of the lower curve.

The above treatment is insufficient if the DC potential difference between the source and the first grid is more than ten times the amplitude of the RF potential. If the DC potential on grid d, to which the AC potential is added, is lowered by just the amount necessary to reduce the speed of an ion receiving maximum incremental energy back to the same speed as that with which the ion passed grid c, and if the DC potential on grid e were also similarly reduced below that of grid d, then symmetric mass line forms will result for AC potentials up to about half the DC potential.

Typical values of grid voltages for a tube resolving negative atomic ions are: ion source a, 0.0 V DC; grid b, + 10.0 V DC; grid c, + 4.0 DC plus 6 V rms AC; grid d, − 2.0 V DC; grid e (stopping potential applied here), −14.1 V DC; collector f, +20.0 V DC. By appropriately selecting the DC potentials, this spectrometer tube serves equally well for separating positive or negative ions. Although no magnetic field is needed with this tube in separating positive ions, it is helpful to use a small magnetic field perpendicular to the axis of the tube when separating negative ions in order to confine the electrons from the source to the vicinity of the source.

Peterlin (1955) derived the resolving power of the three-grid system of the Bennett-type mass spectrometer. In the first approximation, each ion passing the system gains energy

$$\Delta W = \frac{e}{\alpha}\left[F(\theta + \alpha) + F(\theta - \alpha) - 2F(\theta)\right]$$
$$= e \cdot f(\alpha,\theta) \qquad \text{(IV-11)}$$

with

$$F(x) = \int^{x} U(\omega t)\, d(\omega t)$$

This is in the field of the high-frequency signal whose potential is U on the inner grid. The following notations are adapted:

$$\alpha = \omega S/v$$

$$\theta = \omega t_o$$

$$\omega = 2\pi\nu$$

where ν is the frequency of the RF signal, t_o is the time when the ion passes the inner grid, S is the grid separation, v is the mean ion velocity, and e is the ion charge.

With the usual signal shape $U(-x) = -U(x)$, the conditions for a maximum increase in energy as a function of α are

$$\alpha_m U(\alpha_m) = \int_0^{\alpha_m} U(x)dx \tag{IV-12}$$

and

$$\theta_m = 0 \tag{IV-13}$$

The favored ions passing the inner grid at $t_0 = 0$ gain the maximum energy

$$\Delta W_m = 2eU(\alpha_m) \tag{IV-14}$$

In the neighborhood of the maximum, the energy increase may be written as

$$
\begin{aligned}
\frac{\Delta W}{e} &= f_m + \frac{1}{2}(\alpha - \alpha_m)^2 \left(\frac{\partial^2 f}{\partial \alpha^2}\right)_m + \frac{1}{2}\theta^2 \left(\frac{\partial^2 f}{\partial \theta^2}\right)_m + \cdots \\
&= 2U(\alpha_m) + (\alpha - \alpha_m)^2 \frac{U'(\alpha_m)}{\alpha_m} + \frac{\theta^2 [U'(\alpha_m) - U'(0)]}{\alpha_m} + \cdots \\
&= \frac{\Delta W_m}{e} \left\{ 1 + (\alpha - \alpha_m)^2 \frac{U'(\alpha_m)}{2\alpha_m U(\alpha_m)} \right. \\
&\qquad\qquad \left. + \frac{\theta^2 [U'(\alpha_m) - U'(0)]}{2\alpha_m U(\alpha_m)} + \cdots \right\}
\end{aligned}
\tag{IV-15}
$$

where $U'(\alpha_m) < 0$ and $U'(\alpha_m) - U'(0) < 0$. Then the resolution R of the system equals

$$
\begin{aligned}
R &= \frac{m}{\delta m} = \frac{\alpha_m}{2(\alpha - \alpha_m)} = \frac{1}{2}\left(-\frac{\alpha_m U'(\alpha_m)}{\Delta V}\right)^{1/2} \\
&= \rho \left(\frac{U_0}{\Delta V}\right)^{1/2}
\end{aligned}
\tag{IV-16}
$$

The value of α_m corresponds to the transit angle for the case of full width of a peak. U_0 is the amplitude of the signal. ΔV is the difference between the maximum total energy per unit charge of ions leaving the three-grid stage and the stopping voltage. The resolving power increases proportionately to the negative slope of the signal at the moment the ion is entering or leaving the grid system. The resolving power also increases proportionately with the transit time of the ions.

Dekleva and Peterlin (1955) suggested that the three-grid Bennett-type system would have higher resolution if a saw-tooth high-frequency signal as generated in a system shown in Fig. IV-6 were used. Experimental difficulties, however, did not allow them

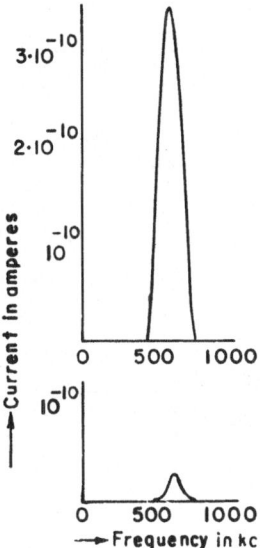

After W. H. Bennett, *J. Appl. Phys.* **21**, 146, (1950).

Fig. IV-5. Ion currents for the Hg⁺ for single stage (Bennett).

to justify completely their statement, but they were able to investigate the influence of the superimposed second harmonics on the resolution of the spectrometer. They applied two signals

$$U_a = U_o \sin \omega t$$

$$U_b = U_o (\sin \omega t - \sin 2\omega t) \qquad \text{(IV-17)}$$

to the RF grid in an experiment. The other parameters chosen were, $U_o \approx 12.3$ V; $U_{eff} = 8$ V; $S =$ intergrid distance ≈ 3 mm; accelerating bias in the ion source was 82 V; the RF frequency was 1.5 Mcps with mass 40. The outcome of the experiment was that

$$\frac{R_b}{R_a} = 2.65$$

where R corresponds to resolution, whereas theory predicted

$$\frac{R_b}{R_a} = 2.243$$

The resolution in the case of pure sine wave voltage V_a may be made equal to that with V_b if the amplitude of the simple sine

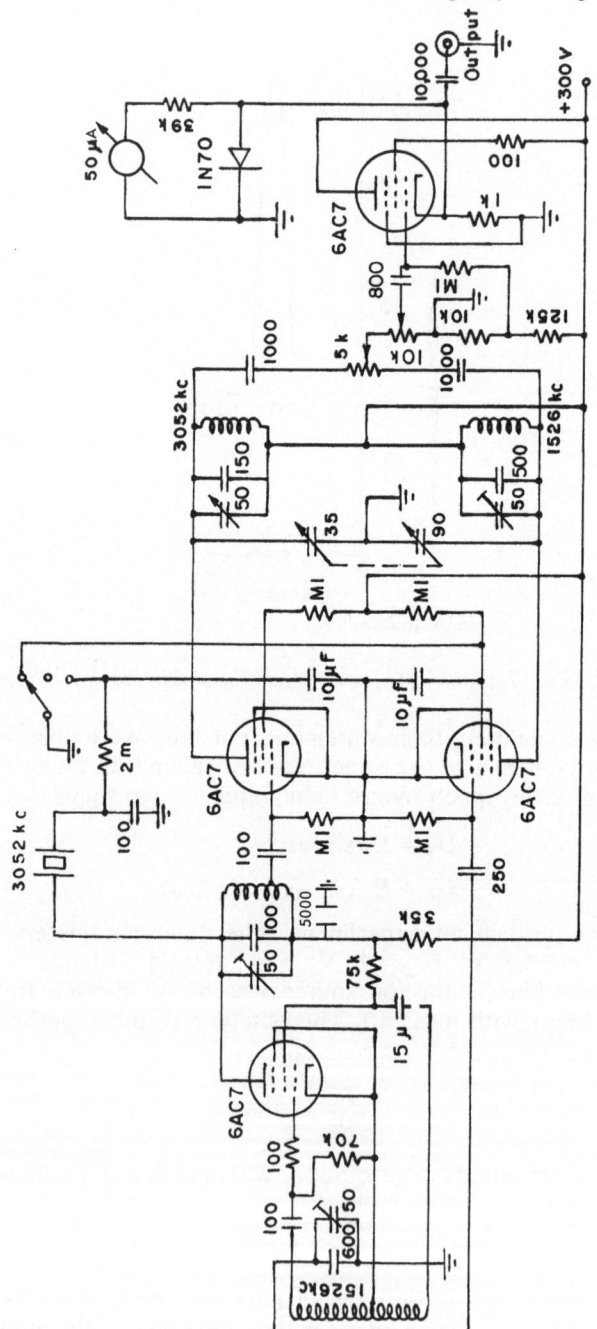

After J. Dekleva and A. Peterlin, *Rev. Sci. Instr.* **26**, 399, (1955).

Fig. IV-6. Dekleva and Peterlin sawtooth RF signal generator.

function is multiplied by 5.03 [= $(2.243)^2$]; i.e., by applying a signal $U_o = 62\,V$. However, such large signals change the condition of the original single stage where the amplitude of the signal must be kept well below the accelerating bias. Therefore, the RF signals cannot be increased enough to replace adequately the effect of the superimposed second harmonics.

b. Two-Stage Tube

The two-stage tube of Bennett is shown in Fig. IV-7. In this case the grids b, c, and d function as the first stage while grids e, f, and g function as the second stage. The same RF potential is applied to grids c and f, in addition to the two unequal DC potentials whose difference provides the compensation. As before, the potential V accelerates the ions from the source a to the first grid b, and DC potentials of grids c, d, f, and g are similarly adjusted. The grids d and e are kept at the same potential so that the region between them is an equipotential drift space. The stopping grid h and the collecting plate p function as before.

The distance between grids d and e is such that an ion receiving maximum incremental energy from the first stage will reach the initial grid e of the next stage at the right phase of the RF

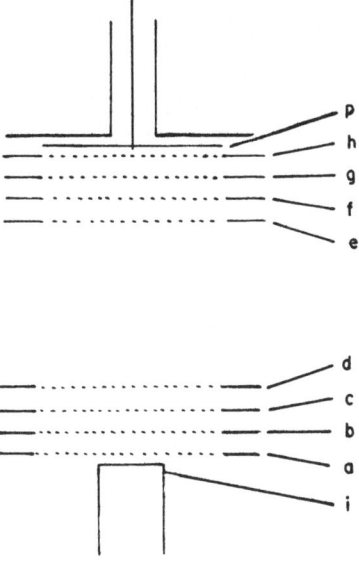

After W. H. Bennett, *J. Appl. Phys.* **21**, 146, (1950).

Fig. IV-7. Two-stage tube (Bennett).

alternating potential to acquire a maximum incremental energy from the second stage also. In order to accomplish this, the RF potential must execute exactly an integral number of cycles while the ion travels from stage to stage, or more exactly, from c to f, and so the distance from d to e must be

$$(2.70\,n - 2)S \qquad\qquad (IV\text{-}18)$$

where n is an integer and S is the intergrid distance.

The first two-stage tube was built with $n = 6$ and it is called a six-cycle two-stage tube. Figure IV-8 represents the incremental energy acquired by an ion from both stages as a function of n, the number of cycles. The envelope of the curve has the same form as in Fig. IV-4 but with twice the ordinate, of course.

In the two-stage tube the selection is much sharper than that obtained with a single-stage tube. High peaks occur for frequencies corresponding to other numbers of cycles between stages than the number 6 for the principal peak, and much of the sensitivity of the tube is lost in increasing the stopping potential to cut out the adjacent peaks.

c. Three-Stage Tube

In order to retain a large sensitivity and at the same time produce only a single peak for each mass analyzed, a third stage is introduced. The three stages must be spaced so that the RF alternating potential on the middle grid of each stage executes exactly an integral number of cycles while the ion acquiring maximum incremental energy travels from stage to stage, i.e., from e to h and from h to k, as shown in Fig. IV-9. These integral numbers are so selected to avoid any appreciable overlap of the har-

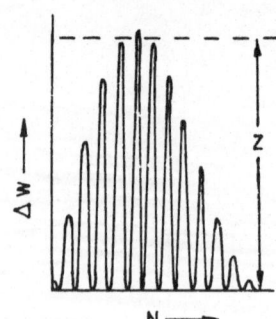

After W. H. Bennett, *J. Appl. Phys.* **21**, 146, (1950).

Fig. IV-8. ΔW vs. N for the two-stage tube (Bennett).

monic peaks of the two pairs of stages when considered as two-stage combinations. As illustrated in Fig. IV-10, when the integral numbers are so selected, more of the principal peak can be left above the blocking potential Z, than with either single-stage or two-stage tubes and a corresponding increase in sensitivity is effected.

Figure IV-9 represents a 9 to 7 cycle three-stage tube which has two different kinds of ion sources. One source utilizes a pencil of electrons, a', which passes at right angles to the axis of the tube. Ions produced along this electron beam are forced to drift toward the grid c by a small electric field between b and c. An alternate source of ions uses electrons from filament a. In this case, the ions produced are accelerated toward grid b.

A DC potential V accelerates the positive ions emerging through c toward and through d. The grids d, e, and f act as the first stage; g, h, and i act as the second stage; j, k, and l represent

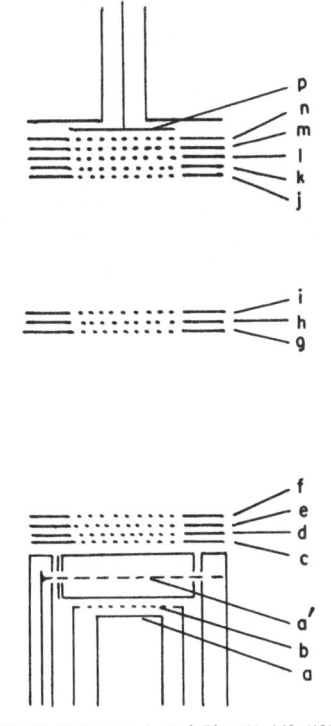

After W. H. Bennett, *J. Appl. Phys.* **21**, 148, (1950).

Fig. IV-9. 9–7 Cycle three-stage tube (Bennett).

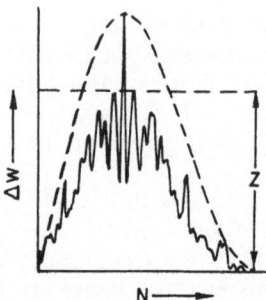

After W. H. Bennett, *J. Appl. Phys.* **21**, 148, (1950).

Fig. IV-10. ΔW vs. N for the three-stage tube (Bennett).

the third stage. The only grids receiving the radio-frequency potential are e, h, and k. A stopping potential is applied between the
collecting plate p and the grid c. Grid m is held at the same potential as the collector and the grid n is held at a negative potential
whose value is more than that applied to any other electrode in
the tube. Grid n will therefore turn back all electrons produced
in the tube between grids d and n. The open area of the grids is
more than 95% and even though 13 grids are used, more than
50% of the beam can pass through to the collector. Grids with
transmission coefficients less than 60% are impractical. The
grids of the first three-stage tube built were spaced at 2 mm in
each stage, which is of the same order of magnitude as the size
of the holes in the grids. Therefore, the grids are far from ideal
in such a case.

Figure IV-11 shows an observation of the spectrum of positive
ions of mercury. Here, harmonic peaks occur at frequencies above

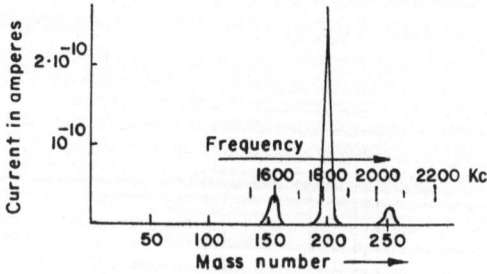

After W. H. Bennett, *J. Appl. Phys.* **21**, 149, (1950).

*Fig. IV-11. Spectrum of the positive ions of mercury, three-stage tube
(Bennett).*

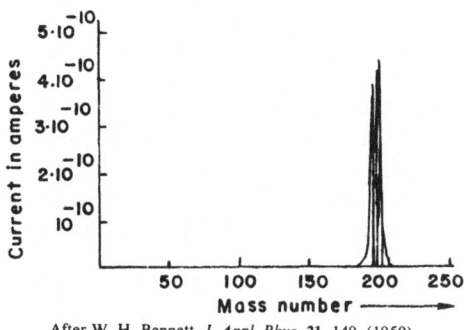

After W. H. Bennett, *J. Appl. Phys.* **21**, 149, (1950).

Fig. IV-12. Finer spectrum of the positive ions of mercury (Bennett).

and below the principal peak which are about one-eighth greater than and less than, respectively, the frequency of the principal peak. These harmonics are evidently caused by a partial overlap of the first harmonic peaks of the two two-stage combinations whose harmonics occur at frequencies equal to one-ninth and one-seventh of the principal frequency. Increasing the stopping potential eliminates these harmonic peaks, as shown in Fig. IV-12.

From the shape of the curve, this spectrometer should have sufficient resolution for a gas analysis where the masses to be completely separated differ by more than 6%. Such is the case in many ordinary gas analyses of the lighter substances. The frequencies at which the various mass components appear is determined by

$$M = \frac{0.266 \times 10^{12}}{S^2 f^2} V \qquad (IV\text{-}19)$$

The instrument, when built with due attention to grid spacing, does not have to be calibrated with known gases and, in this sense, it is an absolute instrument.

This spectrometer tube is found to operate well at pressures from 10^{-6} to 10^{-4} mm Hg. It should be capable of being baked so that the gases usually found in unbaked tubes do not give strong background mass components in the observations. On the other hand, masses can be resolved at pressures up to values of the pressure where the mean free path of the ions is approximately equal to the length of the tube.

Bennett later developed a rapid-scanning RF mass spectroscope where masses ranging from 10 to 50 are swept twice per second and are presented on an oscilloscope. The RF potential is modulated at 10% at one kilocycle and the AC current received at the collector electrode is amplified with an AC amplifier tuned to the modulation frequency. A fixed radio-frequency is used and the principal DC ion-accelerating voltage is swept from 50 to 250 V twice per second. Because of the inherently large sensitivity of this type of mass spectrometer tube, AC current amplifiers with short enough relaxation constants are utilized to permit the rapid scanning of any mass ranges extending over a factor of five.

Henson revised Bennett's single-stage RF spectrometer, greatly improving the resolution. The revised version avoids many grids of the three-stage Bennett type while giving almost the same performance. The method is to apply a square-wave potential to a single stage, the two halves of which are separated by a constant potential "velocity selection" section. The tube consists of the ion source as in Bennett's tube, and also four grids and a collector plate. For the analysis of positive ions, fixed potentials, negative with respect to the ion source, are applied to the four grids. To grids 2 and 3, connected together, is applied the square-wave form consisting of single complete cycles (positive half followed by negative half) separated by intervals. The spaces between grids, the square-wave frequency, and the intervals between cycles are arranged so that an ion of the selected mass passing through grid 1 at the start of a negative half-cycle passes through grid 2 at the end of the negative half-cycle. During the interval between cycles, the ion travels in the field-free space between grids 2 and 3, then emerges through grid 3 at the start of the next positive half-cycle and through grid 4 at the end of this positive half-cycle. The ion is thus accelerated by the square-wave field for the whole of the time it is between grids 1 and 2 and between 3 and 4. At the collector plate a blocking potential is applied thereby collecting only the most energetic ions.

A heavier ion than the selected one travels too slowly to gain maximum energy in either half-cycle. A lighter ion can gain maximum energy from either half-cycle but not from both. Thus ions of the selected mass gain more energy than any others. Different masses are selected by varying the square-wave frequency and the interval length, while keeping the ratio of the period of one cycle to interval between cycles a constant.

The motion of ions in a square-wave field shows that the discrimination between the ions of the selected mass m (atomic mass units) and the next heavier ion of mass $(m + 1)$ is much larger than that between m and $(m - 1)$. The instrument can be used in the low mass range up to mass 50, e.g., it can be used for stable isotope tracer work with carbon, nitrogen, and oxygen. The steady potentials on the grids may be −200 V and the amplitude of the square wave about 60 V. It is considered that an energy discrimination between neighboring masses of at least 2 eV is necessary and that the period of the square-wave form should not be less than 2 μsec. Under these conditions, the tube length would be about 80 cm and the energy discrimination between m and $(m-1)$ varies between 9.6 eV for $m = 13$ and 2.4 eV for $m = 50$. If it is convenient to produce square-wave form with greater amplitude or shorter period, or if a smaller energy discrimination is found to be sufficient, then the tube could be sufficiently shortened.

d. Discussion of Current Efficiency and Half-Width of Resonance Curve

Dekleva *et al.* (1957) discuss the current efficiency and the half-width of the resonance curve in an RF mass spectrometer with respect to the energy gain in the one-, two-, and three-stage tubes. They calculate and measure the quantities for two different analyzing systems; one with a retarding barrier and the other with a deflecting field. In the former case, their measurements show that the relationship between resolving power (r.p.) and current efficiency is r.p. $\approx E_{max}^{-1/5}$ instead of the more useful theoretical relationship, r.p. $\approx E_{max}^{-1}$. In the latter case, the theoretical and the experimental relationships come out to be the same, namely r.p. $\approx E_{max}^{-1/3}$.

The energy gain for the ion in the first stage is

$$\Delta W = \frac{m(v_1^2 - v_0^2)}{2} = eU_0\eta_1 \qquad \text{(IV-20)}$$

where v_1 is the ion velocity after crossing the last grid of the first stage, $v_0 = (2eU/m)^{1/2}$ is the velocity when entering the first stage, U is the accelerating voltage of the ion source, and U_0 is the RF peak voltage. The quantity $U_0 \sin \omega t$ is generally expressed as u.

The factor η determined by Bennett is given by

$$\eta_1 = \frac{2 \sin^2 \alpha \cos \beta}{\alpha} = \eta_0(\alpha) \cos \beta \qquad \text{(IV-21)}$$

where β is the phase angle as the ion crosses the middle grid, $\alpha = s\omega/2v_0$ = semitransit angle, and S is the intergrid distance in each stage. The resonant ions with the maximum η_1 have $\beta = j\pi$ ($j = 0$, $1, 2, \ldots$) and $\alpha_o = 1.166$.

For a two-stage tube,

$$\eta_{II} = \eta_1 + \eta_2 = \eta_o(\alpha)(\cos \beta_1 + \cos \beta_2) \qquad \text{(IV-22)}$$

with

$$\beta_2 = \beta_1 + \frac{r\omega s}{v_o} = \beta_1 + 2r\alpha$$

If $\beta_{II} = \beta_1 + r\alpha$, then

$$\eta_{II} = 2\eta_o(\alpha) \cos(r\alpha) \cos \beta_{II}$$

with the maximum value at $\beta_{II} = l\pi$ ($l = 0,1,2,\ldots$), $\alpha = \alpha_o = 1.166$, and $r\alpha_o = n\pi$ ($n = 0,1,2,\ldots$). The adjacent maxima at $r\alpha = n'\pi$ ($n' = 0,1,2, \ldots$) are much lower than the main maximum.

For three-stage tubes,

$$\eta_{III} = \eta_1 + \eta_2 + \eta_3$$

$$= \eta_o[\cos \beta_1 + \cos(\beta_1 + 2\alpha r) + \cos(\beta_1 + 2\alpha R)]$$

$$= \eta_o(\alpha) \left\{ [2 \cos \alpha R + \cos \alpha (R - 2r)]^2 \right.$$

$$\left. + \sin^2 \alpha (R - 2r) \right\}^{1/2} \cos \beta_{III} \qquad \text{(IV-23)}$$

with the new phase angle given by

$$\beta_{III} = \beta_1 + f(\alpha)$$

where

$$f(\alpha) = \arctan \left[\frac{\sin \alpha (R - 2r)}{2 \cos \alpha R + \cos \alpha (R - 2r)} \right]$$

As in the case of two stages, we have

$$r\alpha_o = n\pi$$

and

$$R\alpha_o = N\pi \qquad \text{(IV-24)}$$

where $N > n$ and $N, n = 0, 1, 2, \ldots$.

e. Five-Stage Tube

Wherry and Karasek (1955) constructed a five-stage (5-9-4-7 cycles) Bennett-type RF mass spectrometer tube. The schematic diagram of the instrument is shown in Fig. IV-13.

The flow of electrons from the heated filament into the collector cage 3 is maintained constant by a control grid and the ions

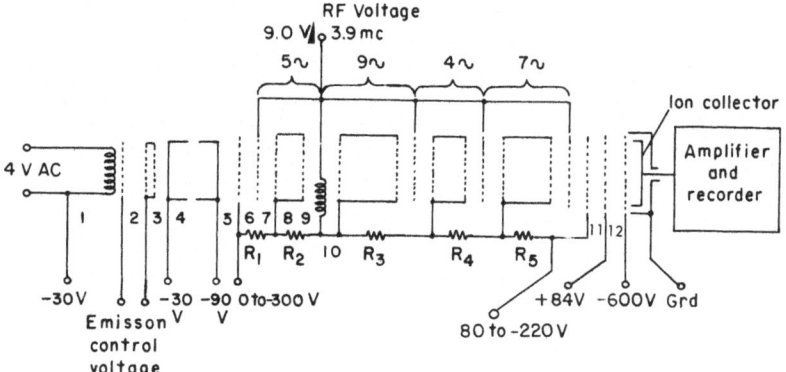

After T. C. Wherry and F. W. Karasek, *J. Appl. Phys.* **26**, 682, (1955).

Fig. IV-13. Schematic diagram of Wherry's five-stage tube.

are formed inside 3 by electron bombardment. The different RF stages are so adjusted that the chosen or resonant ionic mass arrives at grid 11 with maximum energy. A retarding potential on grid 11 will allow only the ion with the highest energy through to the collector. Grid 12 suppresses secondary electrons. Since the RF voltage tends to increase the velocity of the resonant ion, R_1, R_2, R_3, R_4, and R_5 provide stepback voltages to counteract the RF voltage. The resonance condition is

$$M = \frac{0.26 V_a}{s^2 f^2} \tag{IV-25}$$

where V_a is the accelerating voltage, S is the grid spacing in centimeters, f is the frequency in megacycles, and M is the ionic mass. Some of the experimental results obtained with this five-stage tube are shown in Figs. IV-14 through IV-18.

Using this tube, the mass spectra of hydrocarbons from methane through pentane have been determined. A linear mass-scale which follows exactly the equation $M = 0.192 V_a$ is obtained. The electronic circuits required to achieve a reasonable stability are relatively simple. Figure IV-19 represents the mass spectrum of n-butane.

3. REDHEAD'S RADIO-FREQUENCY MASS SPECTROMETER

a. Introduction

Redhead in 1952 developed a linear RF mass spectrometer in which a high velocity ion beam traverses an RF electrode system

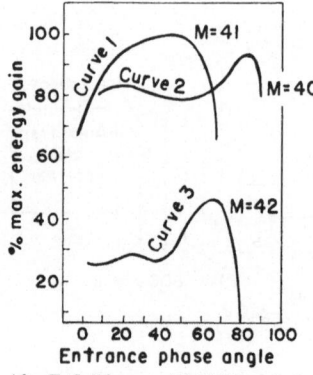

Fig. IV-14. Percent energy gain of ions through the mass selector. The optimum entrance angle for mass 41 is 46.43°; therefore, mass 41 is resonant.

After T. C. Wherry and F. W. Karasek, *J. Appl. Phys.* **26**, 683, (1955).

Fig. IV-15. Calculated values of mass number vs. ion sensitivity, showing discrimination against lighter masses.

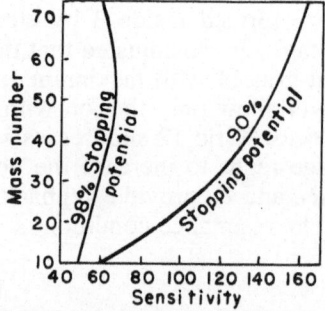

After T. C. Wherry and F. W. Karasek, *J. Appl. Phys.* **26**, 684, (1955).

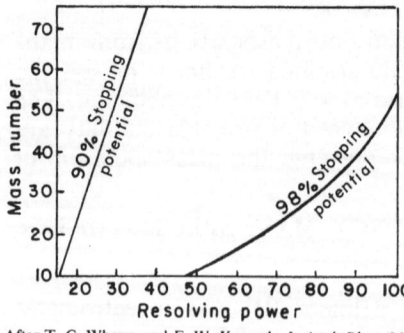

Fig. IV-16. Calculated values of mass number vs. resolving power. Again, the discrimination against lighter masses may be noted.

After T. C. Wherry and F. W. Karasek, *J. Appl. Phys.* **26**, 684, (1955).

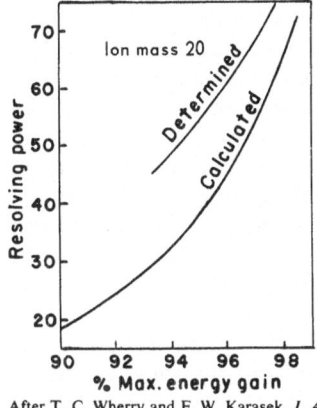

Fig. IV-17. Calculated vs. determined resolving power values at
ion mass of 20. Resolving power
calculated using Raleigh criterion
M/ΔM where ΔM is the peak width
at half the height.

After T. C. Wherry and F. W. Karasek, *J. Appl. Phys.* **26**, 685, (1955).

Fig. IV-18. Calculated vs. determined sensitivity values at an ion
mass of 20.

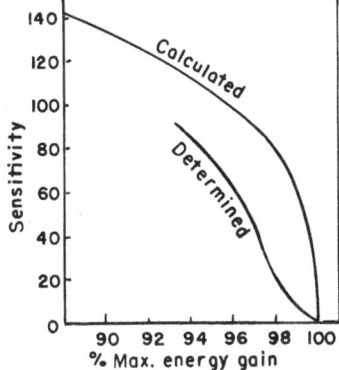

After T. C. Wherry and F. W. Karasek, *J. Appl. Phys.* **26**, 685, (1955).

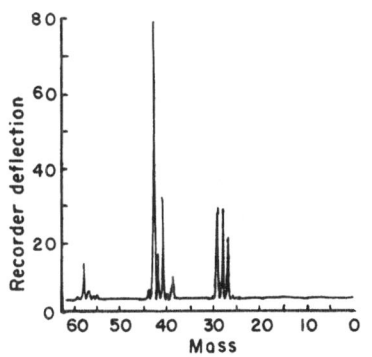

Fig. IV-19. Mass spectrum
of n-butane.

After T. C. Wherry and F. W. Karasek, *J. Appl. Phys.* **26**, 685, (1955).

similar to that of a linear accelerator. Therefore the energy gain of the ions is a sharply selective function of their e/m ratio. Following the RF system, a retarding field energy analyzer is used to select the most energetic ions, i.e., those with a particular e/m ratio.

A monoenergetic beam of ions is accelerated by a DC voltage into an RF electrode system consisting of a series of plane parallel equidistant grids. Alternate grids are grounded and an RF voltage which is small in comparison with the DC accelerating voltage is applied to the remaining grids. If the transit time of an ion between any two grids is half of the RF period, then this ion will remain synchronized with the RF field. The energy gain of the synchronous ion is a function of its entrance phase angle with respect to the RF field and the ions emerging from the RF system encounter a DC retarding-potential barrier which rejects all but the most energetic ions. Thus only the synchronous ion is able to reach the collector. Then,

$$\frac{e}{m} \approx \frac{V_o}{S^2 f^2} \qquad \text{(IV-26)}$$

According to Bennett, energy increment ΔW is given by

$$\Delta W = \Delta \left(\tfrac{1}{2} m v^2 \right) = v \Delta \, (mv)$$

$$= v \int F \, dt \qquad \text{(IV-27)}$$

where v is the initial velocity and F is the force applied to the ion by the RF field. The electric field E_n in the gap between the nth and the $(n + 1)$th grids is

$$E_n = \frac{(-1)^n V}{S f(\omega t + \theta)} \qquad \text{(IV-28)}$$

where V is the peak amplitude of the RF voltage, $f(\omega t + \theta)$ is the form of the applied voltage, θ is the phase of the RF when the ion arrives at the first grid at time $t = 0$, and ω is the angular frequency of the RF voltage. Therefore, the energy gained by an ion traversing a system with N gaps is

$$\Delta W = V \frac{ev}{S} \sum_{n=1}^{N} (-1)^{n+1} \int_{[(n-1)s]/v}^{ns/v} f \, (\omega t + \theta) \, dt \qquad \text{(IV-29)}$$

and the gap transit angle is given by

$$\alpha = \frac{S\omega}{v} = S\omega \left[\frac{m}{2eV_o} \right]^{1/2}$$

If we define $g(\omega t + \theta) = \int f(\omega t + \theta) \, d(\omega t)$, then equation (IV-29) becomes

$$\Delta W = \frac{eV}{\alpha} \sum_{n=1}^{N} (-1)^{n+1} \left[g(\omega t + \theta) \right]_{(n-1)\,\alpha/\omega}^{n\,\alpha/\omega} \qquad \text{(IV-30)}$$

Therefore, if N is even,

$$\Delta W = \frac{eV}{\alpha} \left\{ 2 \sum_{n=0}^{N} \left[(-1)^{n+1} g(n\alpha + \theta) \right] + g(\theta) + g(N\alpha + \theta) \right\}$$
$$\text{(IV-31)}$$

b. Sinusoidal Waveform

   i. Energy Increment Versus Transit Angle. If $g(\omega t + \theta)$ is sinusoidal and is given by $-\cos(\omega t + \theta)$, then equation (IV-31) becomes

$$\Delta W = \frac{eV}{\alpha} \left\{ 2 \sum_{n=0}^{N} \cos(n\alpha + \theta) - 4 \sum_{n=0}^{N/2-1} \cos(2n\alpha + \alpha + \theta) \right.$$
$$\left. - \left[\cos\theta + \cos(N\alpha + \theta) \right] \right\} \qquad \text{(IV-32)}$$

Using the identity

$$\sum_{n=0}^{Q} \cos(A + nB) = \frac{\cos(A + Q/2B) \sin[(Q + 1)B/2]}{\sin(B/2)} \qquad \text{(IV-33)}$$

one obtains equation (IV-32) in the form

$$\Delta W = \frac{2eV}{\alpha} \left\{ \cos\left(\theta + \frac{N\alpha}{2}\right) \frac{\sin[(N + 1)\alpha/2]}{\sin(\alpha/2)} \right.$$
$$\left. - 2 \frac{\cos[\alpha + \theta + (N/2 - 1)\alpha]}{\sin\alpha \cdot [\sin(n\,\alpha/2)]^{-1}} - \cos\left(\frac{N\alpha}{2} + \theta\right) \cos\left(\frac{N\alpha}{2}\right) \right\}$$
$$\text{(IV-34)}$$

The above equation reduces to

$$\Delta W = -\frac{2eV}{\alpha} \cos\left(\frac{N\alpha}{2} + \theta\right) \tan\frac{\alpha}{2} \sin\frac{N\alpha}{2} \qquad \text{(IV-35)}$$

when $N\alpha/2 + \theta = p\pi$, where p is an integer.

The ions arriving at the middle of the RF electrode system in phase or out of phase with the RF voltage gain or lose maximum energy. This maximum energy is

$$\Delta W_m = \frac{2eV}{\alpha}\left| \tan\frac{\alpha}{2} \sin\frac{N\alpha}{2}\right| \qquad \text{(IV-36)}$$

or

$$\Delta W_m(\pi) = \frac{2NeV}{\pi}$$

for $\alpha \approx \pi$. Therefore the energy increment as a function of the transit angle may be expressed as

$$\frac{\Delta W_m(\alpha)}{\Delta W_m(\pi)} = \pi\left[\frac{\tan(\alpha/2)\sin(N\alpha/2)}{N\alpha}\right] \qquad \text{(IV-37)}$$

Equation (IV-37) is plotted in Fig. IV-20 for the case where $N = 20$.

The second highest maximum of the curve occurs near $\alpha = 3\pi$, and

$$\frac{\Delta W_m(3\pi)}{\Delta W_m(\pi)} = \frac{1}{3} \qquad \text{(IV-38)}$$

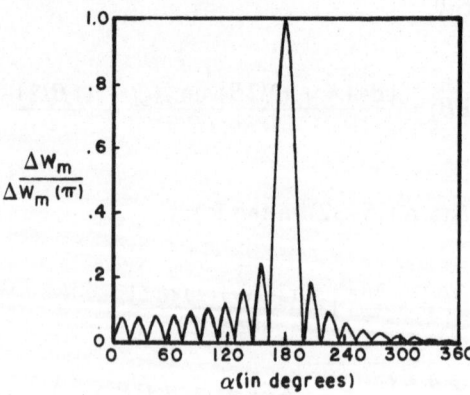

After P. A. Redhead and C. R. Crowell, *J. Appl. Phys.* **24**, 333, (1953).

Fig. IV-20. $\Delta W_m/\Delta W_m(\pi)$ as a function of α for $N = 20$ (sine wave).

Except for the interval $\pi - (2\pi/N) < \alpha < \pi + (2\pi/N)$ when $N \geq 6$, the maxima occurring when $\sin(N\,\alpha/2)$ approaches unity are smaller than the maximum near $\alpha = 3\pi$, since

$$\Delta W_m < \Delta W_m(\pi)\frac{1}{1 - (2/N)}\tan(\pi/2 - \pi/N) \qquad \text{(IV-39)}$$

for α between 0 and 2π. Thus ions having a transit angle other than in the interval $\pi - (2\pi/N) < \alpha < \pi + (2\pi/N)$ may be eliminated by rejecting ions which gain less than $\frac{1}{3}$ of the maximum possible energy.

ii. Resolving Power. The resolving power R of a mass spectrometer may be defined as

$$R = \frac{m_p}{m_2 - m_1}$$

where m_p is the mass of the ion for which the spectrometer is most sensitive and m_1 and m_2 are the mass of the ions which are just rejected. In terms of transit angles corresponding to these ions

$$R = \frac{\alpha_p{}^2}{\alpha_2{}^2 - \alpha_1{}^2} \qquad \text{(IV-41)}$$

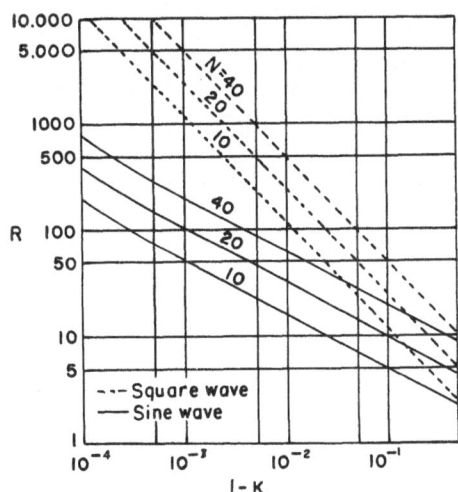

After P. A. Redhead and C. R. Crowell, *J. Appl. Phys.* **24**, 334, (1953).

Fig. IV-21. R vs. (1 − K) for various values of N. Both sine waves and square waves are considered.

If $\alpha \sim \pi$ and $\tan (\alpha/2) \sim 2/(\pi - \alpha)$, then from equation (IV-37)

$$\frac{\Delta W_m}{\Delta W_m(\pi)} = \frac{\sin (N\delta/2)}{N\delta/2} \qquad \text{(IV-42)}$$

where $\delta = (\pi - \alpha)$. If the potential energy barrier is a fraction K of the maximum energy gain, the limiting values of $\delta \, (= \delta_L)$ for which ions can pass over the barrier are those satisfying the condition

$$K = \frac{\sin (N\delta_L/2)}{N\delta_L/2} \qquad \text{(IV-43)}$$

From equations (IV-41) and (IV-43),

$$K = \frac{\sin (N\pi/8\,R)}{N\pi/8\,R} \qquad \text{(IV-44)}$$

Figure IV-21 depicts R against $(1 - K)$ for values of $N = 10, 20,$ and 40.

iii. Current Efficiency. The spectrometer cannot select all ions of one particular mass from a sample since the entrance phase of an ion must satisfy the condition,

$$\frac{\Delta W_m}{\Delta W_m(\pi)} \cos \Delta\theta \geq K \qquad \text{(IV-45)}$$

if the ion is to surmount the potential barrier and reach the collector.

The phase difference from the optimum entrance phase is given by $\Delta\theta$. During one complete cycle the RF system and retarding grid act as a gate which allows the ions to pass when $|\Delta\theta| \leq \Delta\theta_L$. The collector current is therefore pulsed at a radio frequency and the pulse duration is equal to $\Delta\theta_L/\pi$ of an RF cycle. If I_o is the initial current of ions of a given mass and I_s is the average collector current, then

$$I_s = I_o \left(\frac{\Delta\theta_L}{\pi} \right) \qquad \text{(IV-46)}$$

and the current efficiency is given by

$$E_i = \frac{I_s}{I_o} = \frac{\Delta\theta_L}{\pi} \qquad \text{(IV-47)}$$

Eliminating $\Delta\theta_L$ from equations (IV-45) and (IV-47), one obtains

$$\frac{\Delta W_m}{\Delta W_m(\pi)} \cos [\pi E_i] = K \qquad \text{(IV-48)}$$

The maximum current efficiency occurs when $\alpha = \pi$. From equations (IV-44) and (IV-48),

$$\cos{[\pi E_{i\ max}]} = \frac{\sin{(N\pi/8R)}}{N\pi/8R} \qquad (IV-49)$$

In practice $K \approx 1$, resulting in high resolving power and low current efficiency. An approximation to equation (IV-49) is

$$RE_{i\ max} \simeq N/8\sqrt{3} \qquad (IV-50)$$

Better resolving power can therefore be obtained only at the expense of current efficiency, but both can be controlled over a wide range by the retarder grid potential.

c. Square Wave Form

i. Energy Increment Versus Transit Angle. If $g(\omega t + \theta)$ is not sinusoidal but instead a square wave form, then

$$V = \sum_r V_r \sin{(\omega_r t + \theta_r)} \qquad (IV-51)$$

and

$$\frac{\Delta W_m}{\Delta W_m(\pi)} \approx \frac{1 - 2\,|\Delta\theta|/\pi}{1 - \delta/\pi} \qquad (IV-52)$$

where $\delta = (\pi - \alpha)$ and $\theta = N\delta/2 + \Delta\theta$. Equation (IV-52) is plotted in Fig. IV-22 for $N = 20$.

ii. Current Efficiency. Letting $\delta_p = (2/N)\Delta\theta_{L\ max}$,

$$E_{i\ max} = \frac{\Delta\theta_{L\ max}}{\pi} = \frac{1}{2}(1 - K)\frac{1}{1 - (K/N)} \qquad (IV-53)$$

This occurs when $\alpha = \pi - \delta_p$ where $\delta_p = \pi[(1 - K)/(N - K)]$. Also, δ_p is related to α_p, the transit angle of the ion for which the spectrometer is most sensitive.

iii. Resolving Power. Ions will fail to reach the collector when $\Delta\theta = 0$. If $\delta = \delta_1$ or δ_2, where $\delta_1 = \pi - \alpha_1$, $\delta_2 = \pi - \alpha_2$, $\delta = \pi - \alpha_p$, α_1 and α_2 correspond to transit angles of ions of masses m_1 and m_2 which are just rejected and α_p corresponds to the transit angle of the ion for which the spectrometer is most sensitive. Therefore

$$\frac{\Delta W_m}{\Delta W_m(\pi)} = K = \frac{1 - (N\delta_1/2\pi)}{1 - (\delta_1/\pi)} = \frac{1 - (N\delta_2/2\pi)}{1 - (\delta_2/\pi)} \qquad (IV-54)$$

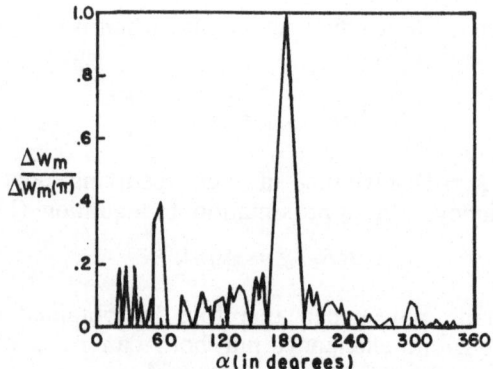

After P. A. Redhead and C. R. Crowell, *J. Appl. Phys.* 24, 335, (1953).

Fig. IV-22. $\Delta W_m/\Delta W_m(\pi)$ vs. α for $N = 20$ (square wave).

Then from equations (IV-41), (IV-53), and (IV-54),

$$R = \frac{1}{8(1-K)}\left(\frac{N-1}{N-K}\right)^2 \frac{(N^2-4K^2)^2}{N(N^2-4K)} \qquad \text{(IV-55)}$$

For large N,

$$R \approx \frac{N}{8(1-K)} \qquad \text{(IV-56)}$$

Multiplying equation (IV-55) by equation (IV-53),

$$RE_{i\ max} \approx \frac{1}{16}\frac{(N-1)^2}{(N-K)^3} \cdot \frac{(N^2-4K^2)^2}{(N^2-4K)} \qquad \text{(IV-57)}$$

For large N,

$$RE_{i\ max} \approx \frac{N}{16}$$

which is approximately the same for the case of a sinusoidal wave [equation (IV-50)].

The energy increment of an ion is proportional to its charge, hence the potential barrier will be a smaller fraction of the peak energy for multiply charged ions than for singly charged ions. Signal peaks due to multiply charged ions should therefore have a higher current efficiency and a lower resolving power than peaks due to singly charged ions. By raising the retarding potential barrier, all ions other than multiply charged ions can be rejected.

For the same value of K and N, considerably higher resolving power can be obtained with square waves than with sine waves.

However, the theoretical maximum for the resolving power cannot be obtained in practice owing to the energy spread of ions from the ion source.

4. AXIAL-FIELD RADIO-FREQUENCY MASS SPECTROMETER

Vorsin *et al.* in 1959 published a comprehensive treatment for an axial-field RF mass spectrometer having $(p + 1)$ stages with p drift spaces. The entry grids of adjacent stages are separated by distances $l_1, l_2, \ldots, l_p$. Each stage consists of $(n + 1)$ grids with n gaps and the gap width is S. A voltage of the form $V_o \sin (\omega t + \theta)$ is applied to the appropriate grids and the incremental energy $\Delta W (\alpha, \theta)$ received by an ion entering the analyzer at $t = 0$ with phase θ and velocity v is shown to be

$$\Delta W (\alpha,\theta) = - 2eV_o \frac{\tan (\alpha/2) \sin (N \alpha/2)}{\alpha} \cdot \left\{ \cos \left(\theta + \frac{N\alpha}{2} \right) \right.$$

$$+ \cos \left[\theta + \left(\gamma_1 + \frac{N}{2} \right) \alpha \right] + \ldots$$

$$\left. + \cos \left[\theta + \left(\gamma_1 + \cdots + \gamma_p + \frac{N}{2} \right) \alpha \right] \right\} \qquad \text{(IV-58)}$$

where $\alpha = S\omega/v$ and $\gamma_1 = l_1/S, \gamma_2 = l_2/S, \ldots, \gamma_p = l_p/S$.

For ΔW to attain a maximum value, it is necessary that

$$\frac{d}{d\alpha} \left[\frac{\tan (\alpha/2) \sin (N \alpha/2)}{\alpha} \right] = 0 \qquad \text{(IV-59)}$$

which is satisfied if

$$\sin \alpha \sin \left(\frac{N\alpha}{2} \right) - \alpha \left[\sin \left(\frac{N\alpha}{2} \right) + \frac{N}{2} \sin \alpha \cos \left(\frac{N\alpha}{2} \right) \right] = 0$$
$$\text{(IV-60)}$$

The maximum value of ΔW corresponds to a tuning parameter α_o given by the principal solution of equation (IV-60), e.g., if $N = 2$, then $\alpha_o = 2.34$, if $N = 4$, then $\alpha_o = 2.93$, etc.

It is also necessary that

$$\theta_o = h\pi - \frac{N}{2} \alpha_o \qquad \text{(IV-61)}$$

and $\alpha_o\gamma_1 = 2\pi n_1, \alpha_o\gamma_2 = 2\pi n_2, \ldots, \alpha_o\gamma_p = 2\pi n_p$ where $h, n_1, n_2, \ldots, n_p$ are integers. The value θ_o defines the optimum entry phase for a synchronous ion and $n_1, n_2, \ldots, n_p$ represent the drift space in number of cycles.

Since the velocity of a synchronous ion may be written

$$v = \frac{S\omega}{\alpha_o} = \left(\frac{2eV}{m} \right)^{1/2} \tag{IV-62}$$

where V is the beam voltage of the ions entering the analyzer, the tuning condition takes the form

$$M = \frac{2NeV\alpha_o^2}{S^2\omega^2} \tag{IV-63}$$

Figure IV-23 represents ΔW as a function of α for the simplest possible analyzer (single stage with three grids) where ΔW_{max} corresponds to $\alpha_o = 2.34$. The secondary maximum in the figure refers to the energy gained by harmonically synchronous ions corresponding to the tuning parameter α_1.

Based on the maximum energy gain considerations, the resolution $M/\Delta M$ is given by

$$\frac{M}{\Delta M} = \frac{\alpha_o}{4\sqrt{2}} \left[\frac{N^2}{12(1-k)} + \left(\frac{2\pi}{\alpha_o} \right)^2 \frac{\begin{Bmatrix} n_1^2 + n_2^2 + \cdots + n_p^2 \\ + (n_1 + n_2)^2 + \cdots + (n_{p-1} + n_p)^2 + \\ \cdots + (n_1 + \cdots + n_p)^2 \end{Bmatrix}}{(p+1)^2(1-k)} \right] \tag{IV-64}$$

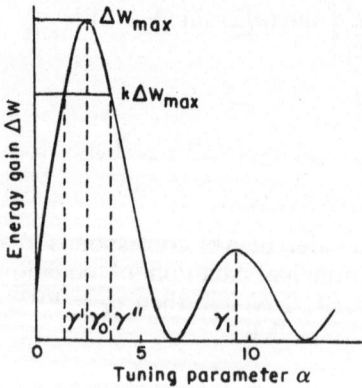

Fig. IV-23. ΔW vs. tuning parameter α for a single-stage RF tube.

and the current efficiency E_i is given by

$$E_i = \frac{\sqrt{2(1-k)}}{\pi} \tag{IV-65}$$

where k is a proportional fraction whereby no ions below an energy level $k\Delta W_{max}$ will continue on to the detector.

If k is set at a typical value of 0.9, a maximum resolution of 24 is obtained when the cycle numbers are 3 and 13. If the grids are transparent, then the current efficiency of the analyzer would be 0.14. The net current efficiency $E_{i\ eff}$ is given by

$$E_{i\ eff} = E_i (T)^{n+1} \tag{IV-66}$$

where T, the grid transmission factor, is 0.95 in this example, $n = 9$, and $E_i = 0.14$. Therefore, $E_{i\ eff} = 0.084$.

Vorsin *et al.* also determined that for a voltage of the form $V_o (\sin \omega t - \sin 2\omega t)$ applied to the Bennett analyzer with a retarder constant 0.9, the number of cycles can be as high as 6 and 22 with no harmonic peaks with a resolution around 40. Figure IV-24 represents the schematic of the radio-frequency mass analyzer.

5. ENERGY ANALYZERS

Even though the principle of energy analyzers has been considered and applied to many cases of mass analysis in the earlier sections, it is desirable to discuss in detail the different types of energy analyzers separately. The two main types of energy analyzers are (a) the retarding-barrier type and (b) the deflecting-field type.

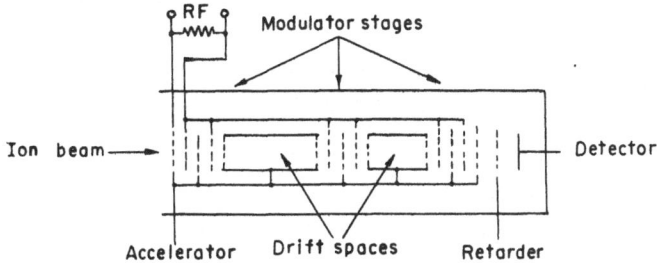

After J. B. Farmer (Chapter on "Types of Mass Spectrometers"), from *Mass Spectrometry* edited by C. A. McDowell, copyright © 1963 by McGraw-Hill, Inc.

Fig. IV-24. RF mass analyzer.

a. Energy Analyzer with Retarding Barrier

The DC retarding potential barrier on grids 11 and 12 shown in Fig. IV-25 rejects all but the most energetic or resonant ions. Such an energy analyzer has been utilized by Bennett and many other experimenters.

After crossing the retarding barrier U_r, the ions with energy gain $eU_o\eta$, have the residual energy eU_{Re}, where

$$eU_{Re} = eU + eU_o\eta - eU - eU_r \qquad \text{(IV-66a)}$$

Even though current efficiency, resolving power, and certain characteristics of RF analyzers have been discussed at considerable length in the previous chapters, it is once again felt essential to discuss the same characteristics of the different energy analyzers with a little more generality.

i. Current Efficiency. If the residual energy is positive for β between the limits β_1 and β_2 (where β is the phase angle as the ion crosses the middle grid), the collector current of the mass m_o is

$$I(\alpha) = \frac{I_o(\beta_2 - \beta_1)}{2\pi} = I_oE_i \qquad \text{(IV-67)}$$

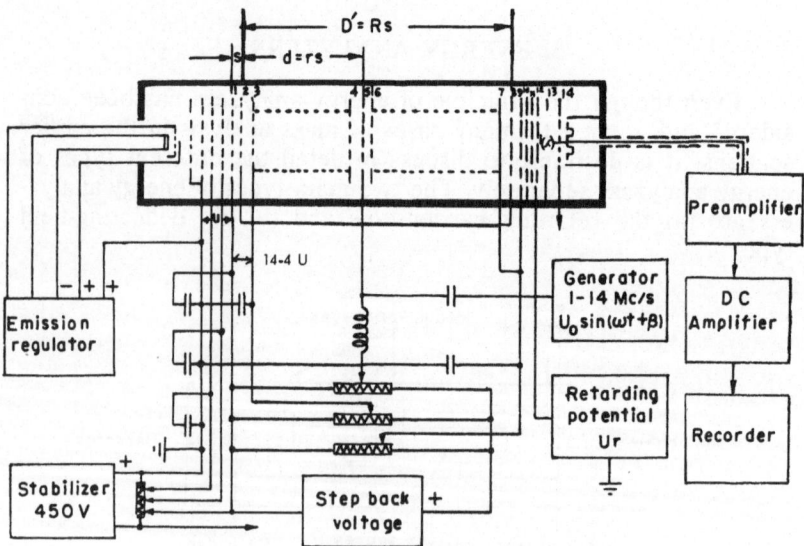

After J. Dekleva and M. Ribaric, *Rev. Sci. Instr.* **28**, 366, (1957).

Fig. IV-25. Block diagram of an RF mass spectrometer with retarding grids.

where I_o is the initial ion current and E_i is the current efficiency. In a three-stage tube, which is the only case of practical importance, the current efficiency is

$$E_i = \frac{1}{\pi} \cos^{-1} \left\{ \frac{3\eta_{3r}}{\eta_o(\alpha)} \left[\left\{ 2 \cos \alpha R + \cos \alpha (R - 2r) \right\}^2 \right. \right.$$

$$\left. \left. + \sin^2 \alpha(R - 2r) \right]^{1/2} \right\} \qquad \text{(IV-68)}$$

where $\eta_{3r} = U_r/3U_o$.

For small values of E_i, equation (IV-68) reduces to

$$E_i(\alpha) = 0.45 \left[1 - \frac{3\eta_{3r}}{\eta(\alpha)} \left\{ [2 \cos \alpha R + \cos \alpha(R - 2r)]^2 \right. \right.$$

$$\left. \left. + \sin^2 \alpha(R - 2r) \right\}^{1/2} \right]^{1/2} \qquad \text{(IV-69)}$$

After expanding in powers of $\Delta\alpha = (\alpha - \alpha_o)$, and after retaining only terms up to $(\Delta\alpha)^2$, the above equation takes the form

$$E_i(\alpha) = 0.45 \left[1 - \left(\frac{\eta_{3r}}{\eta_{o\ max}} \right) \left\{ 1 + \frac{1}{2} \left[\frac{\eta_{\alpha\alpha}}{\eta_{o\ max}} \right. \right. \right.$$

$$\left. \left. \left. + \frac{8}{9} (R^2 + r^2 - rR) \right] \right\} (\Delta\alpha)^2 \right]^{1/2} \qquad \text{(IV-70)}$$

where $\eta_{\alpha\alpha} = \partial\eta^2/\partial\alpha^2$. The maximum current efficiency for a sinusoidal signal is

$$E_{i\ max} = E_i(\alpha_o) = 0.45 \left[1 - \frac{0.23\ U_r}{U_o} \right]^{1/2} \qquad \text{(IV-71)}$$

and it occurs at the same $\alpha = \alpha_o$ as for the maximum energy gain. In general,

$$E_{i\ max} = C \left[1 - \frac{0.23\ U_r}{U_o} \right]^m \qquad \text{(IV-72)}$$

where the value of C is almost impossible to measure since it varies with the instrument used for its determination. Measurements show that $m = \frac{5}{2}$ instead of $\frac{1}{2}$ and this discrepancy has been attributed to the inhomogeneity of the analyzer system.

ii. Resolving Power. Two neighboring masses m and $m + \Delta m$ are considered as resolved if the distance (Δw or ΔU) between the peak maxima surpasses or at least equals the half-height width ($2dw$ or $2dU$ or $2d\alpha = 2\alpha_{1/2}$) of the resonance curve. From this definition, it follows

$$\left| \frac{\Delta m}{m} \right| \geq \frac{4\alpha_{1/2}}{\alpha_o} \tag{IV-73}$$

and resolving power is given by

$$\text{r.p.} = \frac{m}{\Delta m} = \frac{\alpha_o}{4\alpha_{1/2}}$$
$$= 0.107 \, E_{i\,max}^{-1} \cdot h \tag{IV-74}$$

For one stage,

$$h = \left(\frac{|\eta_{\alpha\alpha}|}{\eta_{o\,max}} \right)^{1/2} \tag{IV-75}$$

For two stages,

$$h = \left(\frac{|\eta_{\alpha\alpha}|}{\eta_{o\,max} + r^2} \right)^{1/2} \tag{IV-76}$$

For three stages,

$$h = \left[\frac{|\eta_{\alpha\alpha}|}{\eta_{o\,max} + \,^8/_9 \, (R^2 + r^2 - rR)} \right]^{1/2} \tag{IV-77}$$

where $\eta_{\alpha\alpha} = \partial \eta^2 / \partial \alpha^2$. Since $|\eta_{\alpha\alpha}|/\eta_{o\,max}$ is small compared with r^2 and R^2, the resolving power for a tube with more than one stage depends more upon the distance between the stages than on the single-stage term $|\eta_{\alpha\alpha}|/\eta_{o\,max}$. In general,

$$\text{r.p.} = C' \left[1 - 0.23 \frac{U_r}{U_o} \right]^{m'} \tag{IV-78}$$

where $m' = -\,^1/_2$ from both theory and experiment. Thus, since

$$E_{i\,max} \approx \left[1 - 0.23 \frac{U_r}{U_o} \right]^{5/2} \tag{IV-79}$$

and

$$\text{r.p.} \approx \left[1 - 0.23 \frac{U_r}{U_o} \right]^{-1/2} \tag{IV-80}$$

the two can be combined to give

$$r.p. \approx E_{i\,max}^{-1/5} \tag{IV-81}$$

This equation was discussed earlier, and shows that the instrument is not as useful as one which could be fitted to the theoretical relation, r.p. $\approx E_{i\,max}^{-1}$, because $E_{i\,max} < 1$.

b. Energy Analyzer with Deflecting Field

A three-stage tube with an essentially different analyzer, as shown in Fig. IV-26, was built by Vorsin *et al.* This analyzer is similar to the one used by Harrower for the energy analysis of electrons. In this instrument, the ions, after crossing the third stage, enter a homogeneous electric field inclined by 45° to the direction of the moving ions. In the absence of an RF voltage, the ions reach the collector if the deflecting voltage equals the accelerating voltage ($U_D = U$).

For a given distance x between the entrance slit and the intersection of the deviated ion path in the lower condenser plate, the relation between the change in the deflecting voltage ΔU_D and the change in the accelerating voltage ΔU is given by

$$x = \frac{2d(U + \Delta U)}{(U_D - \Delta U_D)} \qquad \text{(IV-82)}$$

The analyzing power of this deflecting system (defined by the energy of the ions, eU_F, impinging on the collector slit) may be described by

$$U_F = U + U_W\xi + U_M + U_o\eta_n(\alpha)\cos\beta_n \qquad \text{(IV-83)}$$

where U and $U_o\eta_n(\alpha)\cos\beta_n$ are known from the equation

$$eU_{Re} = eU + eU_o\eta - eU - eU_r \qquad \text{(IV-84)}$$

The geometric width of the ion beam is given by $U_W\xi$, $0 < \xi < 1$. U_M gives the displacement of the ion beam beyond the collector in order that the resonant ions, with a maximum loss of energy

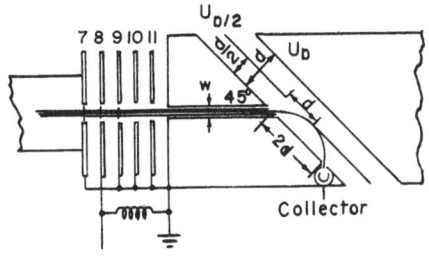

After J. Dekleva and M. Ribaric, *Rev. Sci. Instr.* **28**, 368, (1957).

Fig. IV-26. Analyzer of the RF spectrometer with deflecting field.

only, fall into the collector itself, due to the width of the collector slit.

i. Current Efficiency. The current efficiency is given by

$$E_i' = \int_0^1 \int_{\beta_1}^{\beta_2} d\xi \, d\beta \qquad \text{(IV-85)}$$

where β_1 and β_2 are calculated by putting $U_F = 0$ in equation (IV-83). In this way, the partial current efficiency is

$$E_i'(\xi) = \frac{1}{\pi} \cos^{-1}\left(\frac{U_M - U_W + U_W\xi}{U_o\eta_n(\alpha)} \right) \qquad \text{(IV-86)}$$

when $E_i'(\xi)$ is small, one obtains

$$E_i'(\xi) = 0.45 \left(1 - \frac{U_M - U_W + U_W\xi}{U_o\eta_n(\alpha)} \right)^{1/2} \qquad \text{(IV-87)}$$

On integrating equation (IV-87) with respect to ξ from 0 to ξ_1, where ξ_1 is given by the relation $E_i'(\xi_1) = 0$, the current efficiency is

$$E_i' = 0.3 \, U_o\eta_n(\alpha) \left[\frac{1 - (U_M - U_W)/U_o\eta_n(\alpha)}{U_W} \right]^{3/2} \qquad \text{(IV-88)}$$

and

$$E_{i\,max}' = 0.3 \, U_o\eta_n(\alpha) \left[\frac{1 - \eta_M/\eta_{o\,max}}{U_W} \right]^{3/2} \qquad \text{(IV-89)}$$

where

$$\eta_M = (U_M - U_W)/nU_o \qquad \text{(IV-90)}$$

and n is the number of stages. The agreement between calculated and measured values of current efficiency, $E_{i\,max}'$, is very good.

ii. Resolving Power. By means of equation (IV-90), when $E_i'(\alpha_{o\,n}) = 0$, one obtains

$$\alpha_{o\,n} = 1.42 \, h^{-1} \left[1 - \frac{\eta_M}{\eta_{o\,max}} \right]^{1/2} \qquad \text{(IV-91)}$$

The relation between the current efficiency and peak width results from

$$E_i' = 0.0037 \left[\frac{nU_o}{U_W} \right] h^3 \cdot (\text{r.p.})^{-3} \qquad \text{(IV-92)}$$

where r.p. as before designates resolving power. This relation is substantiated by both theory and experiment.

c. Comparison of the Retarding-Barrier and Deflecting-Field Type Analyzers

In summary, the two basic differences between the two analyzers are: (1) In the retarding-barrier type, there is a linear dependence of the current efficiency on the peak width, while in the deflecting-field type, the dependency of current efficiency on peak width is cubic and so is highly disadvantageous. (2) In the retarding-barrier type, the relation between the resolving power and $E_{i\ max}$ does not depend on the amplitude of the RF voltage, while in the deflecting-field type, the relation between the resolving power and the maximum current efficiency depends on the relation nU_o/U_W. Smaller values of U_W produce favorable results. Because of this, collimation is necessary; the advantage of the RF spectrometers lies in the fact that they avoid the need for collimation.

6. RF PROBE BASED ON THE LINEAR ACCELERATOR PRINCIPLE

In 1955, Boyd and Morris described in detail an RF probe for the mass analysis of ion concentrations. This probe can readily be moved in and out of a discharge tube and can distinguish between ions (positive or negative) of various masses. With this instrument, it is possible to make probe studies of a particular ionic species in the presence of others.

Basically the instrument is a very small 12-stage linear accelerator which discriminates in favor of ions of a particular e/m ratio passing through its sampling orifice. Because of the sampling efficiency, ease in mobility, and the absence of a magnetic field in this instrument, it has certain pronounced advantages over the magnetic method of analysis. This probe is suitable for use in discharges at pressures of 1 mm Hg and higher.

By arranging a system of fields as in the linear accelerator, the diameter of the probe is kept small and a system results in which mass selection is based upon the dependence of ion energy gain on the time of flight. An initial DC accelerating field converts the mass spectrum into velocity spectrum. Since the acceleration due to the radio-frequency fields affects the time of flight, the radio frequency must not be permitted to supply energy to the ions compared to that gained in their initial acceleration. Certain advantages in resolution and sensitivity result from the use of shaped wave forms.

The ions traveling at sub-multiple speeds are rejected by arranging that the time an ion spends in the RF field is an appreciable part of the cycle. Such ions will then experience reversals of field which will greatly reduce their energy gain. The arrangement of grids is shown in Fig. IV-27.

The plane platinum electrode a, which is about 3 mm in diameter, has a small orifice in it. In order to obtain good differential pumping for the highest pressures used, a hole of about 50 μ in diameter is used. In discharges at lower pressures, a considerably larger hole may be employed, thereby giving greater sensitivity. The orifice carries a fine wire grid (14 mm^{-1} pitch, 0.013 mm diameter tungsten wire) to prevent penetration of the accelerating potential through the orifice into the discharge and to give a uniform accelerating field. Experiment has verified that the orifice electrode operates as a Langmuir probe. It may be at space potential or a few volts positive or negative relative to it.

The accelerator electrode b is a grid (pitch and wire diameter same as before) carrying an accelerating DC potential V of between 500 and 1000 V relative to the space potential in front of the orifice. This voltage is large compared with the energy of the ions entering the orifice and so gives rise to a velocity spectrum

$$v = \left(\frac{2Ve}{m} \right)^{1/2}$$

which is dependent almost entirely on the charge-to-mass ratio (e/m) for the ions and not significantly on their initial velocity.

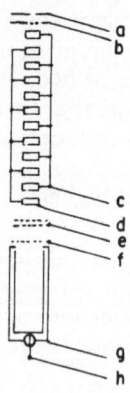

After R. L. F. Boyd and D. Morris, *Proc. Phys. Soc. (London)* **68**, 3, (1955). Courtesy of the Institute of Physics and Physical Society (London).

Fig. IV-27. Schematic diagram of electrode system of an RF mass-spectrometric probe (Boyd).

The radio-frequency electrodes c and d are the elements of a linear accelerator system which acts specifically to accelerate ions of particular velocity, i.e., of a particular e/m ratio. The radio-frequency electrodes are connected to a symmetrical RF oscillator, and have the same DC potential as the accelerator. The amplitude of the applied radio frequency is usually about 5 V so that, although the gain in energy of the ions is sufficient to make discrimination possible, the change in velocity of the flight is so small that the spacing of the electrodes can be kept constant.

The retarder e is a pair of (crossed) grids carrying a retarding potential V_r relative to space potential such that only those ions which have been in resonance with the applied RF have sufficient energy to pass them. The use of two grids effectively eliminates field penetration. The suppressor f is an electrode normally maintained at an accelerating voltage slightly in excess of the initial accelerating potential. Its function is to reject ions of opposite sign to those in the beam, which may have been produced by collisions in the RF region. This is particularly important when working with negative ions, since electrons in the RF region may readily give rise to positive ions by ionization of the residual gas. The collector shield g is a cylinder which shields the collector from electric fields and spurious charges. The collector h (and its shield) is maintained at an attracting potential about 40 V or so above space potential so that particles moving obliquely after passing through the retarder are accelerated into the collector. Electrodes b, c, and d are surrounded by a screen at the same potential as the initial accelerating potential. The screen and all except the radio-frequency and collector electrodes are insulated by silvered ceramic condensers of high permittivity. These condensers form part of the structure.

7. RADIO-FREQUENCY MASS SPECTROMETERS FOR STUDIES OF LIGHT HYDROCARBONS

A radio-frequency mass spectrometer for studies of light hydrocarbons was reported by J. H. Green et al. to the Conference on Mass Spectrometry held in September 1961, at Oxford. The instrument is shown in Fig. IV-28.

The wire mesh used in the grid construction is of knitted tungsten of 0.0005-in. diameter and is stretched over a nickel supporting ring. To ensure that the mesh remains taut, a small nickel disk is placed on top of it and spot-welded to the nickel support below. The mesh is thus tightly sandwiched between the two pieces of nickel. The nickel employed in the construction of the grids is annealed in hydrogen, rolled to a uniform thickness, and after fabri-

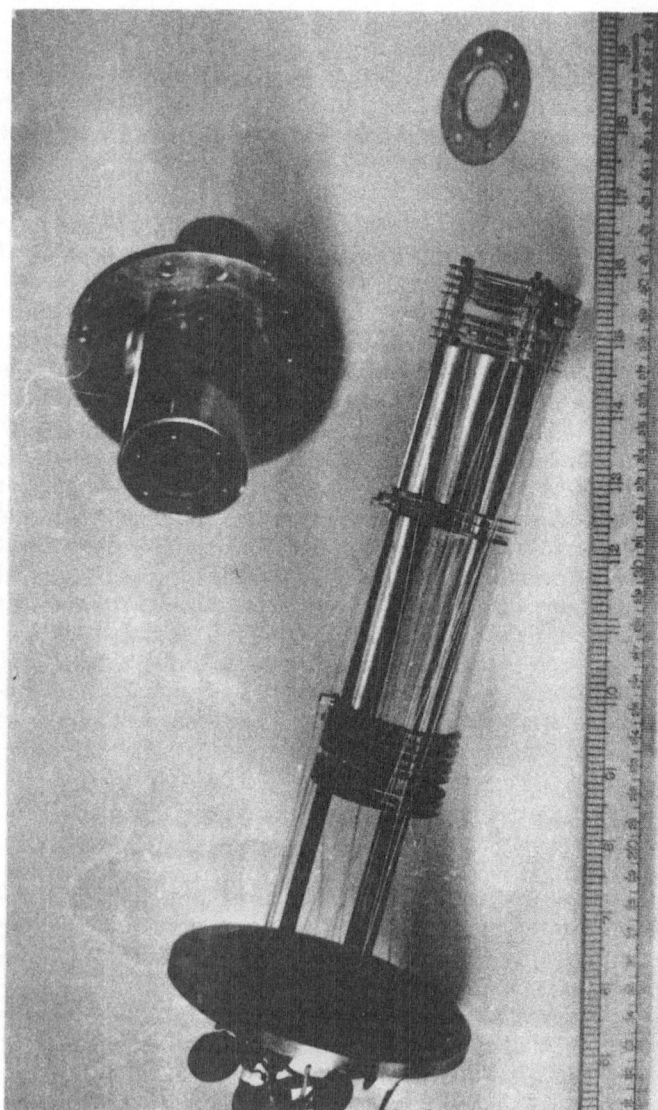

After J. H. Green, *et al*. From *Advances in Mass Spectrometry* – *Vol. 2*, edited by R. M. Elliott, Pergamon Press Book, 1963.

Fig. IV-28. Photograph of the mass spectrometer with different components shown (Green et al., conference at Oxford; hydrocarbons).

cation, selected so that the thickness of the nickel is uniformly 0.0224 ± 0.0001 in. Glass spacers are used to separate the three grids in each individual analyzer stage. The glass spacers are ground to a constant length of 0.1161 ± 0.0001 in. Therefore, the distance between the tungsten mesh for any two consecutive grids in the same stage is 0.1385 ± 0.0002 in. Because the drift space between stages is field-free, the potential on the last grid of one stage is the same as the potential on the first grid of the next stage. For this reason, stainless steel spacers are used to separate any two such grids. The stainless steel spacers are machined and carefully selected so that their lengths are known to within 0.0001 in.

To ensure that the drift spacers are field-free, cylinders of nickel-plated copper are interposed between stages. The resolution of the instrument is greatly improved after these shielding cylinders are added. The collector is a gold-plated Faraday cage which is shielded by a nickel-plated brass cylinder. Ions arriving at the collector are detected and measured by means of a vibrating-reed electrometer. A stainless steel envelope is used for this mass spectrometer and the use of O-rings makes the assembly completely demountable.

There seems to be some confusion regarding the definition of resolving power for these instruments. The resolving powers reported in the literature as 20 to 30 have mostly been based on the ratio of the mass observed to the mass spread at half the peak height. If one uses this criterion, values as high as 80 can be obtained for resolving power. If one defines resolving power as the same ratio at 5% of the peak height instead of at half the peak height, a value of about 30 is obtained. This definition seems to be more practical. A mass spectrum of ethylene obtained using this mass spectrometer is shown in Fig. IV-29.

8. COMPARISON OF RADIO-FREQUENCY AND MAGNETIC MASS SPECTROMETERS

It has been stated repeatedly in the literature that because beam current is not sacrificed by the use of collimating apertures, the RF mass spectrometer has the outstanding advantage over the magnetic type of instrument of having a much higher current efficiency. Now if we consider a simple 180° magnetic mass spectrometer having only two slits, one at the beam entrance and an identical one at the exit, and if the current efficiency is defined as the ratio of the current of ions of a given mass passing through the

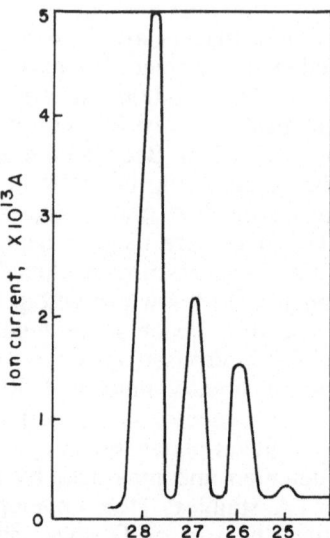

After J. H. Green, *et al.* From *Advances in Mass Spectrometry — Vol. 2*, edited by R. M. Elliott, Pergamon Press Book, 1963.

Fig. IV-29. Mass spectrum of ethylene.

second slit and reaching the collector to that entering through the first slit, this efficiency is unity when the injection energy and magnetic field are correctly adjusted. The collector slit merely prevents ions of unwanted mass from reaching the collector, but does not prevent any of the wanted ions from doing so. This is true as long as the beam passing perpendicularly through the first slit is monoenergetic with small space-charge defocusing. Scattering of the beam by the residual gas in the instrument is also neglected, but it seems unlikely that any of these causes of reduced efficiency is implied in the rather loose assertion above.

It is evident that current efficiency is not suitably defined and should be understood as the ratio of the current of ions of given mass reaching the collector to that *incident* on the first collimating slit. Clearly this can easily be much less than unity when the original beam has a cross section which is much larger than the first slit, so that only a small fraction of ion current can pass into the magnetic analyzer. The current efficiency is therefore a function of the cross section of the original ion beam, and in order to compare the two types of instrument we should consider two identical monoenergetic ion beams free from space-charge defocusing, one incident on the first slit of a simple 180° magnetic mass spectrometer and the other passing without collimation by any aperture into an RF analyzer–collector system. It will be assumed for simplicity that the beams neither diverge nor converge.

It is obvious that variation of the slit widths of the magnetic instrument will control simultaneously both the current efficiency and resolving power, while in the RF type, simultaneous control of current efficiency and resolving power is secured by variation of the retarding voltage V_R. In either case, the current efficiency is functionally related to the resolving power, but the relationship varies between magnetic mass spectrometers and RF mass spectrometers. Therefore, some correspondence principle must be introduced to determine with which magnetic instrument a given RF mass spectrometer may reasonably be compared. The most natural principle to choose seems to be the requirement that the beam path lengths in the two instruments should be equal, since this length is an important consideration in limiting the background gas pressure. The following comparison is based on this assumption.

Suppose, for simplicity, that the ion beam has a cross section $a \times b$ with $a \leqslant b$. The circular beam of diameter D is then approximated with little error by one of square cross section with $a = b = D$. In the case of the magnetic instrument, let the current efficiency be reduced to $E_i = a_c/a$, by the use of an entrance slit $a_c \times b$ and a collimating aperture of the same dimensions. Then for a given beam injection voltage

$$\frac{\delta m_o}{2m_o} = \frac{\delta R}{R} \qquad (\text{IV-93})$$

where m_o is the mass of an ion and R is the radius of curvature of its path. Since $\delta R = a_c/2$ determines the limiting curvature of ion paths reaching the collector, we have

$$\frac{\delta m_o}{m_o} = \frac{a_c}{R} \qquad (\text{IV-94})$$

or, in terms of the beam path length l, and current efficiency E_i,

$$E_i = \frac{1}{\pi} \frac{l}{a} \left| \frac{\delta m_o}{m_o} \right| \qquad (\text{IV-95})$$

For an N-stage periodic analyzer, we have

$$E_i \sim \frac{1}{4\sqrt{3}} N \left| \frac{\delta m_o}{m_o} \right| \qquad (\text{IV-96})$$

The above equation is applicable only when there is no attenuation of the beam by any grids. Now RF analyzers may be divided into two distinct cases.

In the first class, grids are not used and the RF electrodes are hollow cylinders (Boyd, 1955), or in the hypothetical case of rectangular ion beams, they are hollow rectangular boxes. Since a is the minimum width of the beam, it is clear that the length of one stage cannot be made much less than a without introducing a severe reduction of the RF field along the ion beam, and no gridless analyzer seems to have been described in literature in which the stage length is less than a. In this class of analyzers, we can therefore assume that the maximum number of stages which can be accommodated along a length l of the ion beam is rather less than l/a, and considering equation (IV-96), one can easily see that

$$E_i < \frac{1}{4\sqrt{3}} \frac{l}{a} \left| \frac{\delta m_o}{m_o} \right| \quad \text{(gridless)} \quad \text{(IV-97)}$$

It is clear by comparing equations (IV-95) and (IV-97) that at fixed resolving power, the current efficiency of a gridless RF instrument is less than half that of the magnetic spectrometer of equal path length.

The second class of analyzers uses grids as the RF electrodes, and in this case, there is obviously no necessary connection between the minimum width of the beam a and the grid spacing. Such analyzers may therefore have $N \geqslant l/a$, so that

$$E_i \gg \frac{1}{4\sqrt{3}} \frac{l}{a} \left| \frac{\delta m_o}{m_o} \right| \quad \text{(IV-98)}$$

An instrument for which $N \leqslant l/a$ was once suggested by Boyd in 1955 for rocket exploration of the ionosphere. If one assumes that the grids used are of 95% transparency and unaligned, which is certainly the case in the square wave analyzer, then one finds for this instrument

$$E_i \sim \frac{1}{8} \frac{l}{a} \left| \frac{\delta m_o}{m_o} \right| \quad \text{(IV-99)}$$

and for the aperiodic analyzer (Bennett-type)

$$E_i \sim \frac{1}{1.1} \frac{l}{a} \left| \frac{\delta m_o}{m_o} \right| \quad \text{(IV-100)}$$

Comparing equations (IV-98), (IV-99), (IV-100), with (IV-95), it is evident that only the Bennett mass spectrometer gives higher current efficiency than the equivalent magnetic instrument and that the increase in E_i is less than a factor of three. The generally held view that RF mass spectrometers give current efficiencies an order of magnitude greater than magnetic instruments is therefore unsupported.

It is widely believed that when compared with a magnetic instrument, an RF mass spectrometer has relatively poor resolving power. Judging from the comparison of equations given above, this is rather surprising because it would seem that an RF type may be operated to give any required resolving power with current efficiency similar to a 180° magnetic type of equal path length. The equations are, however, based on the assumption of a perfect retarder–collector assembly, but in practice it does not hold.

The instrumental difficulty arises as follows: The retarding voltage V_R is normally derived from a DC supply unrelated to the RF oscillator voltage V, and even with careful feedback regulation of the oscillator, it is difficult to prevent changes of less than 2% in the ratio (V_R/VG_M). Under high resolution conditions, this resolution must be maintained close to the cutoff value $V_R/VG_M = 1$, and small changes in the ratio are enough to cause E_i to vary erratically over quite a large range. From the relations

$$\frac{V_R}{VG_M} = \cos \pi E_i \qquad (IV-101)$$

and

$$E_i = \frac{N}{4\sqrt{3}} \left| \frac{\delta m_o}{m_o} \right|$$

it can be shown that

$$\left| \frac{\delta E_i}{E_i} \right| = \left| \frac{\delta \rho}{\rho} \right| \frac{48}{N^2 \pi^2 (\delta m_o/m_o)^2} \qquad (IV-102)$$

where $\rho = V_R/VG_M$. Equation (IV-102) is plotted in Fig. IV-30.

If we assume that the ion currents must be measured with an accuracy of at least 10%, then $|\delta E_i/E_i| \leq 0.1$, so that we require from Fig. IV-30,

$$\left| \frac{\delta m_o}{m_o} \right| \geq \frac{22}{\pi N} \left(\left| \frac{\delta \rho}{\rho} \right| \right)^{1/2} \qquad (IV-103)$$

The best attainable resolution, in practice, is therefore inversely proportional to N and directly proportional to the square root of the percentage variation in ρ. The second cause for limited

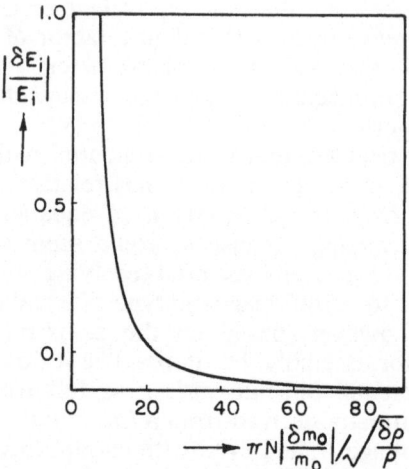

After L. W. Kerr, *J. Electronics* **2**, 196, (1956). Courtesy of Taylor and Francis, Ltd.

Fig. IV-30. $\delta E_i/E_i$ *vs.* $\pi N \left| \dfrac{\delta m_0}{m_0} \right| \times \sqrt{\rho/\delta\rho}$.

resolving power is the retarder–collector assembly. Figure IV-31 depicts the ideal collection efficiency characteristic as a function of ion energy. It also depicts the form obtained in practice when lens effects occur in the retarder assembly.

In most of the energy-gain mass spectrometers, the retarding voltage V_R is applied to the collector and to a screening grid in front of it. The imperfection of the collection-efficiency behavior seems to be due to scattering of the beam in the strong inhomogeneous fields near the grid wires. The relations

$$E_i = \frac{1}{\pi} \cos^{-1} \left| \frac{\sin[(\pi/4)(N\delta m_0/m_0)]}{(\pi/4)(N\delta m_0/m_0)} \right| \qquad \text{(IV-104)}$$

and

$$E_i = \frac{N}{4\sqrt{3}} \left| \frac{\delta m_0}{m_0} \right|$$

yield a new current efficiency curve as shown in Fig. IV-32, in terms of a new parameter P equal to VG_M/V_I, where V_I is the beam injection energy.

The claim that RF mass spectrometers of the energy-gain type have rather limited resolving power seems to be fairly well founded. It is possible, however, that the use of a more carefully

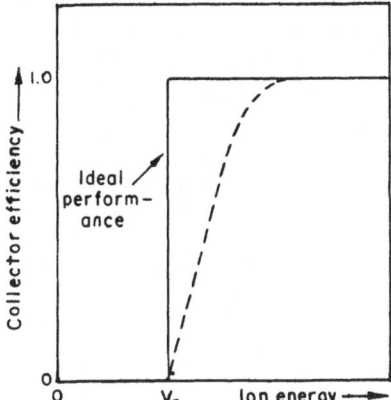

After L. W. Kerr, *J. Electronics* **2**, 197, (1956). Courtesy of Taylor and Francis, Ltd.

Fig. IV-31. Collection efficiency vs. ion energy (Kerr).

designed retarding system and electronic stabilization of the ratio V_R/V would allow one to follow the ideal performance curve more closely.

The advantages of the RF mass spectrometer over its magnetic equivalent are important in the mass analysis of gas discharge ions where the stray magnetic field of the conventional instrument may modify the discharge structure and produce extraneous mass dis-

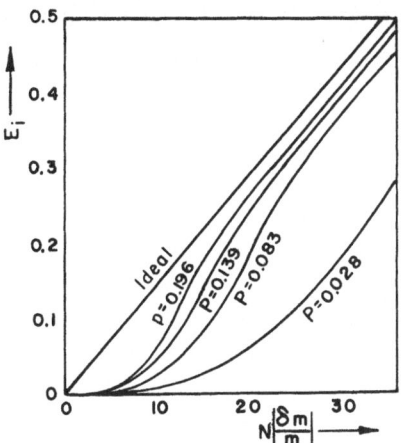

After L. W. Kerr, *J. Electronics* **2**, 197, (1956). Courtesy of Taylor and Francis, Ltd.

Fig. IV-32. Ideal E_i vs. $N\left|\dfrac{\delta m}{m}\right|$ (Kerr).

crimination in the sampling orifice region. For the observation of weakly ionized gases such as the ionized layers of the earth's atmosphere, RF mass spectrometers using grid analyzers allow a very much higher current efficiency than their magnetic equivalents, and there is no limit to the beam cross section and sampling area of such instruments.

Chapter V

Cyclotron-Resonance Instruments

1. INTRODUCTION

In certain types of mass spectrometers, the ions describe circular paths in homogeneous magnetic fields at the cyclotron frequency given by

$$f = \frac{eH}{2\pi MC} \qquad \text{(V-1)}$$

where H is the strength of the magnetic field and M is the ionic mass. Here the mass of the ion is determined by measuring the cyclotron frequency usually in terms of a harmonic. These "cyclotron-resonance" instruments possess high resolution and are designed specifically for the study of atomic masses.

2. HELICAL-PATH MASS SPECTROMETER

The helical-path mass spectrometer was proposed by Goudsmit in 1948 and built by Hays, Richards, and Goudsmit in 1951. In this instrument, the ions from a pulsed source move in helical paths in a magnetic field of large volume. The pitch of the helix is sufficient to allow the ions to miss the source structure as they complete their first revolution. The detector is located directly beneath the source and records the time of arrival of the ions as they complete their nth revolution. The cyclotron frequency associated with the ion in question is determined by measuring the time required to complete one full cycle. As discussed earlier, the cyclotron frequency is given by the relation $f = eH/2\pi MC$.

Since $f \propto 1/M$, the mass scale, when based upon frequency, is linear in this instrument. Thus the mass precision, ΔM, is constant over the entire mass range. In practice, the elapsed time between the second and eleventh revolutions changes by about

10 μsec/mass unit, and since this could be measured to an accuracy of 0.01 μsec, the precision of the mass determinations is approximately 10^{-3} mass units or one MeV. This instrument is particularly useful in the study of heavy atoms and is calibrated by standard hydro and fluoro carbons. In the study of ^{208}Pb and ^{209}Bi, a few of the calibration masses being used are $C_3F_5 (M = 131)$, $C_4F_7 (M = 181)$, and $C_5F_9 (M = 231)$.

The magnetic field is produced by a spherical air core magnet which, because of its extent, is of low intensity and necessitates the use of low-energy ions (25 eV ions for $M = 100$). These low-energy ions are strongly influenced by polarization of the walls of the vacuum chamber. It is not practicable to increase appreciably either the strength or the volume of the magnetic field, the former to reduce this polarization effect and the latter to allow a large number of spirals. Although the instrument has not been fully exploited for these reasons, it has made a timely and significant contribution to our knowledge of atomic masses.

3. TROCHOIDAL PATH RADIO-FREQUENCY SPECTROMETER

The large five-cycle trochoidal mass spectrometer at the National Bureau of Standards was converted to a cyclotron-resonance instrument by Hipple and Sommer in 1953. The ions, after traversing one cycle, pass between a pair of small, closely spaced electrodes to which the RF voltage is applied. If this RF voltage is approximately zero at the time of their arrival, the ions will be undeflected and will continue on a trochoidal path until they reach a second similar pair of electrodes four cycles later. Otherwise, they will be deflected in the direction of the magnetic field. The radio frequency, f', which will permit the undeflected passage of the ions through the second pair of electrodes and hence to the collector, is simply related to the cyclotron or fundamental frequency f as follows

$$f' = \frac{n}{8} f \tag{V-2}$$

where n in this particular case is the number of half-cycles of the radio frequency. The cyclotron frequency, incidentally, is not affected by the presence of the electric field.

In operation, the DC field is first adjusted to bring the desired ions to the collector. Then the RF field is turned on, adjusted, and its frequency measured. A similar procedure is followed for a second ion of neighboring mass, thereby permitting the mass difference to be computed from

$$\frac{M_1}{\Delta M} = \frac{f_2}{\Delta f}$$

or,   (V-3)

$$\frac{M_2}{\Delta M} = \frac{f_1}{\Delta f}$$

A resolution of 1/12,000 at half-maximum could be achieved.

4. OMEGATRON

The omegatron of Hipple, Sommer, and Thomas (1949, 1950) is essentially a small cyclotron in which ions are accelerated if the frequency of the RF accelerating field coincides with the cyclotron frequency. This principle had been employed earlier by Alvarez and Cornog (1939) with an actual cyclotron to demonstrate the existence of ^{3}He.

The maximum radius of curvature of the omegatron is 0.9 cm. The RF field is both uniform and weak, pervading the entire accelerating region, in contrast to the established D-shape characteristic of high-energy accelerators. Ions are created in the central region by bombardment, and spiral out to the collector, making 3000 to 5000 revolutions en route. They are prevented by a DC field from escaping axially. The large number of revolutions results in a sharp mass discrimination, corresponding to a resolution of 1/10,000 for low masses. The mass spectrum is obtained by varying either the frequency or the magnetic field.

Although this instrument has been used to make an important preliminary determination of the hydrogen–deuterium mass difference (Sommer, Thomas, and Hipple, 1951), the space-charge difficulties have interfered with its use with heavier atoms. Its chief application has been in the determination of the proton–moment in nuclear magneton units. This was done by measuring the cyclotron frequency and the nuclear magnetic resonance frequency in the same magnetic field.

Bloch and Jeffries (1950) and Jeffries (1951) have used a small cyclotron to effect a resonant deceleration of ions. Here, 20 keV protons initially following an orbit of 4.25 cm radius reach the central collector after 500 revolutions. As with the omegatron, this equipment has been used to determine accurately the proton moment. H$^+$ ion peaks have been obtained which indicate a resolution of approximately 1/10,000 at half-maximum, and the intention has been expressed of modifying the instrument for use in determining the masses of light atoms.

5. SMITH'S MASS SYNCHROMETER

This instrument, developed by L. G. Smith at the Brookhaven National Laboratory, is shown schematically in Fig. V-1. The mass synchrometer is an outgrowth of Smith's association with the helical path device. Here the ions, while constrained by a uniform magnetic field to follow circular paths, are exposed in the "pulser" to a local modulating field. The slit S_3 is connected to the appropriate source of potential.

In the first model of the synchrometer this potential consisted of negative rectangular pulses of one-microsecond duration, occuring at variable and measurable time intervals. With the first pulse, a group of ions in the neighborhood of S_3 is decelerated sufficiently to miss the source housing and enjoy a free circulation in orbit 2. After these ions have made a number of such revolutions, a second pulse decelerates the group still further with the result that they reach the collector one half-cycle later along orbit 3. When Smith conducted the experiments with this instrument, ions of energy 250 V and mass 28 were observed after completing 90 revolutions. The resolution obtained at half-maximum was 1/24,000 but the intensity reached a very low level.

The design specifications of the mass synchrometer as detailed by Smith (1951) are rather interesting. The instrument was designed to be in a metal chamber in the 2-in. gap of an electromagnet with poles 15 in. in diameter. The diameter of orbit 2 in Fig. V-1 is 10 in. with H = 8200 Oe. At a pressure of approximately 5 × 10^{-6} mm Hg, ions of masses 18, 28, and 44 formed in the residual gas have been observed after 70, 40, and 25 rotations, respectively, between pairs of pulses. The respective values of the resolution

$$\frac{M}{\Delta M} = \frac{nT}{\Delta(nT)} = \frac{L}{2\Delta L} \qquad (\text{V-4})$$

are between 8.5 and 3 × 10^3. The main advantage of the new synchrometer is that its finite size does not limit L and hence the resolution as in the crossed-field instruments.

In the second model of the mass synchrometer (Smith, 1952–1953), S_3 is connected to an RF oscillator. This effects a harmonic modulation of the orbit diameter. It is arranged that one cyclotron revolution takes place in $(n + \frac{1}{2})$ cycles of the RF. Thus, after one

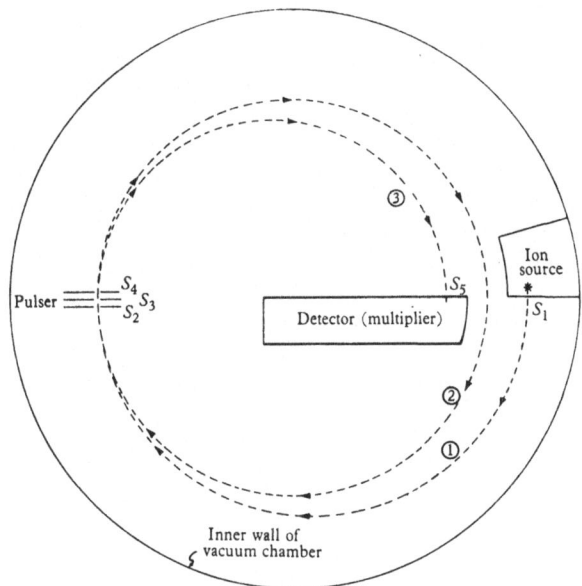

After Lincoln G. Smith, From *Mass Spectroscopy* (H. E. Duckworth), Cambridge University Press, (1958), p. 82.

Fig. V-1. Smith's mass synchrometer.

revolution, the ions reach the pulser in such a phase as to cause a demodulation of the diameter. In this way most of the desired ions are reassembled for detection and the long ion path of the original model is avoided. In practice, the demodulation at the end of one revolution is not as efficient as was hoped, and the actual resolution achieved is considerably below expectations. Also, the background current is large and there is serious interference from "satellites" which have made three or more revolutions (Smith and Damm, 1956).

The latest model of the mass synchrometer (Smith and Damm, 1956) has recently been described in detail. Here the ions which are finally collected are decelerated by the modulator on each of three successive revolutions. In order to select these ions, two slits are selected between S_1 and S_5, say S_6 and S_7, with S_6 being nearer to S_1. After the first deceleration at S_1, the ions pass through S_6, reaching the modulator again in such a phase as to be further decelerated by approximately the same amount. The orbit then passes through S_7. After the third and final deceleration, the orbit passes through S_5 to the detector, which is in this case a Faraday collector plus a DC amplifier rather than an electron multiplier of earlier models. The choice of three revolutions is made to avoid satellites.

The cyclotron frequency is determined in terms of its nth harmonic, given by the frequency of the applied RF voltage corresponding to an ion peak. In a typical case ($M = 32$), n is 107 and the applied frequency is 14.9 Mc. For ion energies of 2500 eV, modulating voltages of 180 to 3200 V rms have been employed. The diameter of the first orbit is 11 in.; this is decreased in three stages, as described earlier, to a final value of about 9 in. The half-width resolution is adjustable electrically between values $\frac{1}{10,000}$ and $\frac{1}{25,000}$ for all mass numbers below 250. In practice, securing satisfactory ion intensity has been a problem.

The mass difference between two ions with masses M_1 and $M_2 = M_1 + \Delta M$ is given by either

$$\frac{M_1}{\Delta M} = \frac{f_2}{\Delta f}$$

or

$$\frac{M_2}{\Delta M} = \frac{f_1}{\Delta f}$$

as has been discussed earlier. The cyclotron frequencies for the two ions in question are given by f_1 and f_2, and $\Delta f = f_1 - f_2$.

The frequency is measured with high precision as has been described by Smith and Damm (1953, 1956). In the first place, the frequency of the RF field is periodically swept, thereby permitting the amplified output of the collector to be presented as a single peak on an oscilloscope. Further, f is displaced on alternate sweeps by an amount Δf, so adjusted (along with amplifier gain DC level and a small fractional increment $\Delta M/M$ in accelerating voltage) that the two peaks on the oscilloscope screen appear coincident. Also on alternate sweeps, a square wave voltage is fed to the second y input of the oscilloscope. This causes one of the peaks to appear as two, vertically displaced with respect to one another, between which the other peak may be accurately centered. This peak-matching technique has made possible some of the very accurate mass comparisons and has been incorporated in modified form by Nier in his double-focusing mass spectrometer.

Chapter VI

Massenfilter as a Mass Spectrometer

1. INTRODUCTION

A mass filter which is able to separate ions of different specific charge, e/m, by means of a high-frequency electric quadrupole field was proposed and tested by Paul, Reinhard, and Zahn (1953–1955). The mass filter appeared to be a mass spectrometer of high resolution and a rather efficient isotope separator. In 1958, Paul *et al.* published a detailed theoretical and experimental analysis of the mass filter.

2. EQUATIONS OF MOTION IN A QUADRUPOLE FIELD

In the mass filter the ions are introduced into a field produced by a potential of the form

$$\Phi = \Phi_o(\alpha x^2 + \beta y^2 + \gamma z^2) \qquad \text{(VI-1)}$$

where $\alpha + \beta + \gamma = 0$, $\alpha = -\beta = 1/r_o^2$, $\gamma = 0$, and $\Phi_o = U + V \cos \omega t$ is the potential applied to the electrodes. Therefore,

$$\Phi = (U + V \cos \omega t)\frac{x^2 - y^2}{r_o^2} \qquad \text{(VI-2)}$$

Figure VI-1 represents the end-on view of a quadrupole mass analyzer. The equations of motion of a singly charged ion are

$$m\ddot{x} + 2e(U + V \cos \omega t)\frac{x}{r_o^2} = 0 \qquad \text{(IV-3)}$$

$$m\ddot{y} - 2e(U + V \cos \omega t)\frac{y}{r_o^2} = 0 \qquad \text{(VI-4)}$$

$$m\ddot{z} = 0 \qquad \text{(VI-5)}$$

119

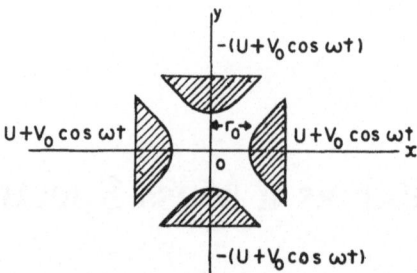

Fig. VI-1. End-on view of quadrupole mass analyzer.

The last equation leads to $\ddot{z}$ = constant. Equations (VI-3) and (VI-4) are known as Mathieu's differential equations and describe the oscillations of an ion under the influence of a periodic force.

Using the following transformations, $\omega t = 2\xi$, $a = 8eU/mr_o^2\omega^2$, and $q = 4eV/mr_o^2\omega^2$, one obtains

$$x'' + (a + 2q \cos 2\xi)x = 0 \qquad (VI-6)$$

and

$$y'' - (a + 2q \cos 2\xi)y = 0 \qquad (VI-7)$$

where the primes denote differentiation with respect to ξ. All solutions to equation (VI-6) may be expressed in the form

$$x = \alpha' e^{\mu\xi} \sum_{-\infty}^{\infty} C_{2s}e^{i2s\xi} + \alpha'' e^{-\mu\xi} \sum_{-\infty}^{\infty} C_{2s}e^{-i2s\xi} \qquad (VI-8)$$

where μ is the characteristic exponent and is a constant.

There are two types of solutions to equation (VI-8): (1) the stable solution where, as ξ approaches infinity, x remains finite for all ξ, and (2) the unstable solution where, as ξ approaches infinity, x approaches infinity. The stability of the solution depends only upon μ, which is a constant and does not contain the initial conditions, as may be seen from the following: (1) if $\mu = i\beta$ and β is not a whole number, then the solutions are stable; (2) if $\mu = i\eta$ and η is a whole number, then the solutions are nontrivial, limited, periodic, and unstable; (3) if μ is complex, the solutions are unstable. By increasing the ratio U/V the stable q-interval, which corresponds to a stable mass interval, may be made so small that only ions of one mass are able to pass through the field. All other ions will traverse unstable paths, strike the electrodes, and be removed. This is the fundamental filtering action of the electric quadrupole field which permits its use as a mass spectrometer.

a. Frequency Spectrum of the Solutions

Inside the stable region where $\mu = i\beta$ and $0 < \beta < 1$, the solution of equation (VI-3) has the form

$$x = \alpha_I \sum_{-\infty}^{\infty} C_{2S} \cos\left(S + \frac{\beta}{2} \right)\omega t + \alpha_{II} \sum_{-\infty}^{\infty} C_{2S} \sin\left(S + \frac{\beta}{2} \right)\omega t \quad \text{(VI-9)}$$

where the initial conditions appear in the constants of integration, α_I and α_{II}, and the fundamental frequency ω_o equals $(\beta/2)\omega$. The solution for y is similar.

b. Initial Conditions and the Maximum Amplitude of Oscillation

It has been assumed that the mass filter distinguishes between stable and unstable ions by the trajectory stability criterion which depends only upon the operating point (a, q) and which is independent of the initial conditions. The maximum amplitude of oscillation, however, depends both upon the operating point and upon the initial conditions. From equation (VI-9) one finds that the greatest possible amplitude is

$$|x_M| = (\alpha_I{}^2 + \alpha_{II}{}^2)^{1/2} \sum_{-\infty}^{\infty} |C_{2S}| \quad \text{(VI-10)}$$

Although the solution, $x(\xi)$, is in general not strictly periodic, the actual amplitude of oscillation may approach x_M after a few oscillations.

In order to compute $\alpha_I(\xi_o, x_o, x_o')$ and $\alpha_{II}(\xi_o, x_o, x_o')$, the fundamental system of solutions $x_I(\xi)$ and $x_{II}(\xi)$ is used. By differentiating

$$x(\xi) = \alpha_I x_I(\xi) + \alpha_{II} x_{II}(\xi) \quad \text{(VI-11)}$$

and introducing the initial values, one obtains

$$|x_M| = \frac{1}{W} \sum_{-\infty}^{\infty} |C_{2S}| \left\{ [x_o x_{II}'(\xi_o) - x_o' x_{II}(\xi_o)]^2 \right.$$

$$\left. + [x_o' x_I(\xi_o) - x_o x_I'(\xi_o)]^2 \right\}^{1/2} \quad \text{(VI-12)}$$

where W is the Wronskian determinant. For a given value of ξ_o, the quantity under the square root sign is a biquadratic form in

x_o and x'_o. If $x_M = r_o$, then equation (VI-12) is an ellipse in the (x_o, x'_o) plane.

Fisher carried out the numerical computations for $a = 0$ and $\beta = 0.2$, 0.5, and 0.8. The maximum amplitude of oscillation was greater than the initial displacement of the ion from the quadrupole axis for all phases of high-frequency voltage at the instant the ion enters the field except at $\xi_o = \pi/2$ where the two are equal. As a result, only a definite cross section of the quadrupole field is available for injecting ions into the field.

3. USE OF MASS FILTERS AS ELECTRIC MASS SPECTROMETERS

a. Amplitude of Vibration

The use of a mass filter as a mass spectrometer requires that it possess a high resolution and that the relative abundances of the individual masses in a mass spectrum can be measured with sufficient accuracy. Higher resolution may be obtained by a proper choice of U/V, thereby the mass line on the stability diagram (see Figs. VI-2 and VI-3) has a steeper slope. In this way the operating point moves closer to the apex of the stability triangle.

For a resolution greater than 70, only the region in the stability diagram bounded by $0.69 < q < 0.71$ and $0.23 < a < 0.24$ is of interest. The apex will therefore lie at $q_{lim} = 0.70600$ and $a_{lim} = 0.23699$. For $q = 0.706$, then β_x and β_y are given by

$$(1 - \beta_x)^2 = \frac{0.23699 - a}{1.93750} \qquad \text{(VI-13)}$$

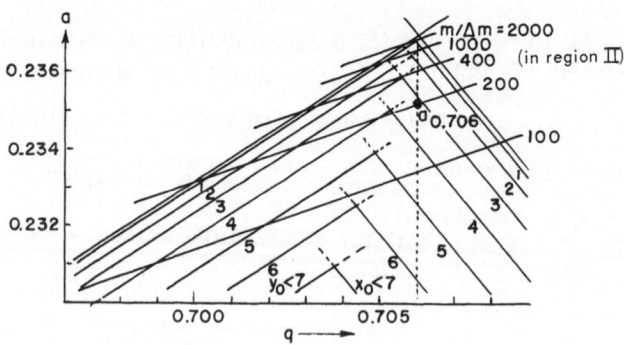

After W. Paul, et al., Z. Physik **152**, 146, (1958). Courtesy of Springer-Verlag, Berlin.

Fig. VI-2. Stability diagram, plot of a versus q.

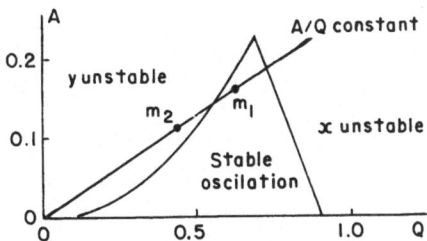

After J. B. Farmer. From *Mass Spectrometry* edited by C. A. McDowell, copyright © 1963 by McGraw-Hill, Inc.

Fig. VI-3. Stability diagram for the quadrupole mass analyzer.

and

$$\beta_y^2 = \frac{0.23699 - a}{0.79375} \tag{VI-14}$$

The expansion coefficients, C_{2S}, are computed using recursion formulas. By using equations (VI-13) and (VI-14), one can obtain the maximum amplitude of stable ions as a function of the initial conditions ξ_o, x_o, x_o', y_o, and y_o'. It can clearly be visualized that x_M and y_M increase inversely proportional to $1 - \beta_x$ and β_y, respectively.

b. Peak Shape and Transmission for Injection Parallel to the Axis

In this case $\dot{x}_o$, $\dot{y}_o = 0$ and x_o, $y_o \neq 0$ with the ratio x_M/x_o a well-defined function of β (see Figs. VI-2 and VI-3). Along the operating point on the mass line in the stable region, the intensity, a function of amplitude, will at first increase proceeding from the first stability limit. If the input aperture is small, a point will finally be reached where all ions corresponding to the stable mass will reach the collector. A transmission of 100% of the ions introduced to the field is then obtained. Approaching the other stability limit an intensity decrease occurs. In the ideal case the peak shape described will be a trapezoid where the plateau indicates the region of 100% transmission. However, this case leads to a maximum intensity independent of a very low resolution. At high resolution the peak shape becomes triangular and the intensity decreases inversely proportional to the resolution.

To increase the probability that a stable ion with $\dot{x}_o$, $\dot{y}_o = 0$ can pass through the field, the initial displacement of the ion from the field axis is restricted by the use of an entrance aperture of suitable diameter D.

c. Resolution

The half-width, ΔM, of the line M is used as a measure of resolution. Operating within region I, where the width of the sides of the peak is small compared to the total width of the line, the half-width is practically the same as the total width. For $q = 0.706$ and indicating the ordinate of the mass line in the stability diagram by $a_{0.706}$, then in region I,

$$\frac{M}{\Delta M} = \frac{0.178}{0.23699 - a_{0.706}} \tag{VI-15}$$

In region II, which is of interest in high resolution, the peaks are triangular and the half-width is approximately equal to one-half of the full width. For $q = 0.706$, then

$$\frac{M}{\Delta M} = \frac{0.357}{0.23699 - a_{0.706}} \tag{IV-16}$$

The resolution is now related to β and hence x_M and y_M through equations (VI-13) and (VI-16). For injection parallel to the axis in region II the maximum amplitude increases only with the square root of the resolution since

$$\frac{x_M}{x_0}, \frac{y_M}{y_0} < 1.8 \left(\frac{M}{\Delta M} \right)^{1/2} \tag{VI-17}$$

However, this increase becomes important when operation at high resolution is desirable. With $x_M, y_M = r_0$, the optimum diameter of the entrance orifice is given by

$$D \approx \frac{r_0}{(M/\Delta M)^{1/2}} \tag{VI-18}$$

If the ions are introduced on the axis with a radial velocity, then

$$\dot{x}_M, \dot{y}_M < 0.16 \, r_0 \omega \left(\frac{\Delta M}{M} \right)^{1/2} \tag{VI-19}$$

It was assumed in deriving equations (VI-15) and (VI-16) that the ion remains a certain length of time in the field. A definite number n of high-frequency periods are therefore required in order that the amplitude of the unstable ion will become large enough so that the ion will be removed. According to Zahn, n is given by

$$n \approx 3.5 \left(\frac{M}{\Delta M} \right)^{1/2} \tag{VI-20}$$

d. Design Criteria of Mass Filters
The high-frequency and DC voltages necessary to place the operating point at the apex of the stable region are

$$V = 7.219 \, Af^2 r_0^2 \qquad \text{(VI-21)}$$

and

$$U = 1.212 \, Af^2 r_0^2 \qquad \text{(VI-22)}$$

where f is the frequency in megacycles, r_0 is in centimeters, and A is the atomic weight of the stabilized ion. Noting that $U/V = a_g/2q_g$ at the apex of the stability region, then $U_g/V_g = 0.1678$. The high-frequency power P required is therefore (in watts),

$$P = 6.5 \times 10^{-4} \frac{CA^2 f^5 r_0^4}{Q} \qquad \text{(VI-23)}$$

where C is the capacitance of the system in picofarads and Q is the figure of merit of the power circuit.
Stable ions introduced along the field axis with radial energy smaller than U_r will be focused where

$$U_r \approx \frac{V}{15} \cdot \frac{M}{\Delta M} \qquad \text{(VI-24)}$$

The maximum allowable accelerating voltage U_β of the ions is

$$U_\beta^{max} \approx 4.2 \times 10^2 f^2 L_m^2 A \frac{\Delta M}{M} \qquad \text{(VI-25)}$$

where L_m is the length of the quadrupole field in meters.
The operation of the mass filter is determined to a great extent by the location of the operating point in the stability diagram, i.e., the parameters a and q must be stabilized about the ratio $2M/\Delta M$. In addition to holding U and V constant and varying f to bring the masses one after another into the stable region, one may hold f constant and vary U and V such that U/V remains a constant. If, for example, in the case of increasing V, the ratio U/V slowly increases, the mass line approaches more closely the limiting value (a_g, q_g) in the stability diagram or may even pass above this point. As a result, the ratio of the measured intensities of the peaks corresponding to two masses no longer gives the true relative abundance of these masses in the original beam. Since, as verified both theoretically and experimentally, the intensity I, in region II, varies as

$$I \sim \frac{\Delta M}{M} \qquad \text{(VI-26)}$$

If I_2^0/I_1^0 is the true relative abundance of the masses A_2 and A_1, and I_2/I_1 is the measured ratio, then

$$\frac{I_2^0/I_1^0 - I_2/I_1}{I_2^0/I_1^0} = \frac{(U/V)A_2 - (U/V)A_1}{U_g/V_g - (U/V)A_1} \approx \frac{A_2^\alpha - A_1^\alpha}{(0.1678/Ck^\alpha) - A_1^\alpha}$$

(VI-27)

where $U \approx CV^{1+\alpha}$, $V = kA$ with $\alpha > 0$. See Fig. VI-4 for the change in resolution and collector ion current as given by the above equations.

The most critical quantity in the entire design is the field radius, r_o, which must be held to within $\pm 10^{-4}$ of its designated value. The hyperbolic surfaces assumed as boundary conditions in the Mathieu differential equations cannot be fabricated with sufficient accuracy. In the actual experiment the hyperbolic surfaces were replaced with rods of circular cross section; the ratio $r_{rod}/r_{field} \approx 1.16$ is the best approximation to the hyperbolic field. The field should be as short as possible because of the difficulty in mounting the quadrupoles with sufficient accuracy. The frequency should be as low as possible since the high-frequency power varies as f^5. For example, if $L = 1$ m, $f = 4$ Mc, and $U_\beta = 75$ V, the power required at mass 16 is approximately 300 W and $|V| \approx 4$ kV. The maximum entrance aperture was about 0.4 mm.

e. Description of the Apparatus

In the massenfilter described by Paul et al., the ions were introduced in an electron-impact ion source. The ion-accelerating voltage need not be stabilized and one can even operate with AC line voltages. A transition field which is not very well known is found in the region between the grounded entrance aperture and the quadrupole field. In order to reduce this field effect upon the ion

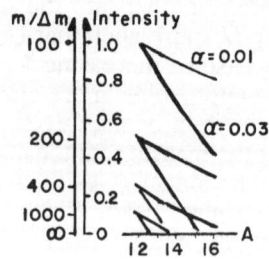

After W. Paul, et al., Z. Physik **152**, 154, (1958). Courtesy of Springer-Verlag, Berlin.

Fig. VI-4. Change in resolution and collector ion current.

paths to a minimum, the ions were introduced through a canal which extended into the rod system. The rods of the quadrupole system were made of brass and were supported at three places by "Ergan" plate rings, 1 cm thick. The collector, which had a large cross-sectional area, was located a few centimeters away from the quadrupole system. The current to the collector was measured by a vibrating-reed amplifier connected to a recorder. For an input resistance of $10^{10}\,\Omega$, a full-scale deflection corresponded to 10^{-10}, 10^{-11}, or 10^{-12} A. The effect of the high-frequency field on the zero of the amplifier could be completely eliminated.

The high frequency was produced by a separately excited push-pull stage, whose tank capacity was the quadrupole system itself. The DC voltage U was obtained directly from the high frequency by means of an inductive voltage divider. An especially difficult problem was realizing a symmetrical field, since both U and V must be symmetrical with respect to the axis. Since an unsymmetrical field has its greatest effect on the initial conditions and does not change the separation of the ions, only the transmission is decreased.

4. EXPERIMENTAL RESULTS

a. Measurements at Low Resolution ($M/\Delta M < 80$)

A wide field application of the mass spectrometer is to determine the relative abundance of the masses in a mass mixture. In order that the ion current measured at the collector of the massenfilter be a true measure of this relative abundance, it must be assumed that there is a well-defined, reproducible relation between the two, as is clearly the case if the transmission is 100%. Figure VI-5 shows a mass spectrum of methane.

Some investigations using electron-impact sources with magnetic mass spectrometers have shown that the measured relative abundance depends upon the geometry of the source, the temperature of the source, and the impressed voltages. Upon eliminating these dependencies in the present experiment, there was an increase in intensity of 9% in 75 min. However, with resolution no systematic change in the relative abundance was detected. Region I is thus suitable throughout its entire extent for the measurement of relative abundances. The precision of the measurement depends upon the constancy of the ion source and detector and not upon the quadrupole field.

b. Effect of Gas Pressure

In contrast to magnetic mass spectrometers it is possible to operate the massenfilter at relatively higher gas pressures. The

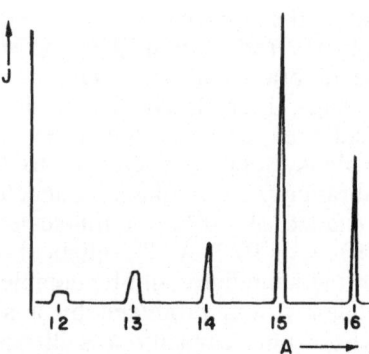

After W. Paul, *et al.*, Z. *Physik* **152**, 154, (1958). Courtesy of Springer-Verlag, Berlin.

Fig. VI-5. Mass spectrum of methane.

properties of the ions which determine whether they are stable or unstable are not changed in the first approximation by a collision with a background gas molecule. Therefore, one would expect that the characteristic pressure-broadening of the peaks obtained with the magnetic mass spectrometer would be smaller in the case of the massenfilter.

c. Resolution

Qualitatively, one can see that the amplitude actually increases more rapidly with radial energy in the y-direction than in the x-direction, that is, the corresponding mass line edge is flatter. However, one must assume a radial energy as large as 1 eV in order to explain the fact that the transmission has fallen below 100% as soon as $M/\Delta M$ becomes greater than 80. This may be caused by the transition between the grounded inlet aperture and the normal quadrupole field. Brubaker and others have pointed out that the quadrupole mass spectrometer is highly suitable for studying the composition of the atmosphere in the vicinity of an artificial satellite because of the insensitivity to the velocities of the incoming molecules or ions.

Finally, the limiting voltage U_g at which the intensity of the mass 15 peak of methane just disappeared was determined. This voltage was found to be 2% greater than the value computed from equation

$$U_g = \frac{0.23699 M \omega^2 r_0^2}{ge} \qquad \text{(VI-28)}$$

Thus, the computed value is in error by about $\pm 0.5\%$ which is attributed to the uncertainty of the proper value for r_o. The measured value of U_g is known to an accuracy of $\pm 0.2\%$.

Since the derivation of the expression for U_g is strictly valid only for hyperbolic electrodes, these results show that the cylindrical-rod arrangement used produces a field gradient which approximates the ideal quadrupole field to a high degree. Qualitatively, any symmetric arrangement of electrodes whose cross section represents a section of an arbitrary quadratic surface will produce an ideal quadrupole field over a definite region surrounding the axis of symmetry. It is not self-evident that, in the case of the rod geometry chosen, the distance of the rod surface to the field axis, which was arbitrarily assumed to be equivalent to r_o, would give such accurate agreement between theory and experiment.

Chapter VII

Applications of Nonmagnetic Mass Spectrometers to Upper Atmosphere Research

1. INTRODUCTION

Investigation of the ionic spectrum and the gaseous composition of the upper atmosphere is of paramount importance for the solution of many geophysical and astrophysical problems. The atmospheric pressure and temperature cannot be measured or determined accurately unless the gaseous composition is very well known. The problem of locating the level of gravitational separation of gases, that is, the level below which the atmosphere is mixed and above which there occurs a regular distribution of the gaseous composition with altitude (relative increase in the amount of light gases with increasing altitude), can be solved only after a detailed investigation of the gaseous composition of the atmosphere. The atmospheric structure is the basic knowledge necessary to tackle any other physical problem associated with the atmosphere.

The problem of the dissipation of the atmosphere into outer space (helium problem) can also be resolved by knowledge of the gaseous make up of high-altitude strata. The problem of solar corpuscular radiation can be clarified considerably when reliable data on the gaseous composition of the atmosphere in the polar regions is obtained. Investigation of the ion composition of the atmosphere may also help to explain the occurrence and existence of ionized layers at different altitudes. Such an explanation becomes possible on the basis of studying the diurnal variation of the ion composition and also the nighttime concentrations.

For the purposes of efficient radio communications, it is extremely important to ascertain the spectrum of the ions and their effective cross sections for collisions with electrons. The effective cross sections of ions are naturally quite different from the cross sections of neutral atoms or molecules. In the region between 30 to 80 km, rocket-borne sample bottles have been successfully employed. It is likely, however, that the collection of usable quantities of gas is limited to 100 km even in spite of using liquid hydrogen to "freeze out" the sample. To add to this difficulty at higher altitudes is the possibility of contamination due to leaks and the possibility of selective absorption and recombination at the walls of the bottle.

Above 100 km, a mass spectrometer can easily be used to provide a continuous record of composition throughout its working range of pressures. Since the data can be telemetered to earth, no recovery of the apparatus is necessary. Mass spectroscopy of the D region is a complicated problem. Extremely rigid requirements and design problems confront the experimenter. Only recently has the mass spectra of D region (Narcisi and Bailey) been obtained. The experimental and engineering problems seem to have been solved with considerable success. Investigation of the E and F regions is fairly easy with rocket-borne mass spectrometers. Satellites are also used to carry the mass spectrometers to provide information about the world-wide distribution of neutral and positive ions. Negative ion spectroscopy is still in an infant stage. In a few years, we may know as much about the negative ion composition of the atmosphere as we today know about the positive ion composition. Rockets and satellites have their own advantages and disadvantages. It is very well known that any measuring instrument perturbs the physical situation and the vehicles which carry the instruments inject considerable error into the measured values of various physical parameters.

One of the decisive advantages of experiments performed using a satellite is the prolonged stay of the measuring apparatus in the layers investigated. Combined with the tremendous speed of the satellite, this feature permits multiple and almost simultaneous observations at points separated by tens of thousands of kilometers. The time difference between investigations made in equatorial zones and polar regions may not exceed 20 to 30 min, and the repeated appearance of the satellite in the same zones makes it possible to trace the time variations of the ion composition. The oblate orbit of a satellite permits measurement of the ion composition at various altitudes above the earth, mainly the E and F regions. The satellite has a characteristic advantage over the rocket. The rocket, upon entering rarefied layers of the atmosphere, evolves large amounts of parasitic gases which surround the rocket with a unique gas cloud. This cloud consists of the air that had filled the rocket prior to its ascent and the rocket has no opportunity to leave the contaminated region. However, a satellite, because of its thoroughly sealed small volume, minimizes the gases evolved from the inside and the prolonged stay of the satellite in the rarefied layers permits a thorough outgassing of the satellite surfaces.

2. PROBLEMS IN THE PERFORMANCE OF SATELLITE EXPERIMENTS

The satellite is located in the regions of the atmosphere so rarefied that the mean free paths of the molecules will reach tens and hundreds of meters. Since the satellite velocity is an order of magnitude higher than the gas kinetic molecular velocities, it is important to consider the perturbation of the medium due to vehicle motion.

It is of primary importance to ascertain if the satellite measures the true ionization. Meteoric ionization provides considerable insight into this problem. Meteor flights through the atmosphere are accompanied by intense ionization of the atmosphere. Meteors have a speed greater than 8 km/sec, of course, and ionization is observed in layers lower than those in which the satellite moves. The effect of satellite motion in the ionosphere can be considered in detail.

To ionize the molecules of the surrounding medium, a body must have a minimum energy of 15 eV. The satellite itself does not have this energy, and therefore cannot ionize the gas molecules in its path directly. But molecules undergoing elastic collision with the satellite have enough energy for ionization by impact. This is extremely important. It is also necessary that the nature of the collisions be ascertained, and experimental results predict that the collisions are inelastic. Also, all meteoric theories presuppose inelastic collisions between the meteor and the gas molecules. However, assuming that the collisions are elastic, the number of molecules reflected from the satellite surface in time ΔT is

$$\Delta N = N_o S_o v_o \Delta T \qquad \text{(VIII-1)}$$

where N_o is the number of molecules per unit volume in the medium surrounding the satellite, S_o is the area of the satellite bombarded by the molecules, and v_o is the satellite velocity.

The rebounded molecules are redirected toward the forward half of a sphere centered about the point of initial impact and the radius of this sphere is λ, the mean free path. The first collision of the rebounded molecule takes place on the surface of this sphere. Within the time ΔT, the satellite moves a distance $\Delta\lambda$ and consequently, the collision between the "fast" and the "slow" molecules in the surrounding medium takes place in a layer bounded by λ and $\lambda + \Delta\lambda$. The volume of a hemispherical layer having these radii is

$$V = 2\pi\left(\lambda^2\Delta\lambda + \lambda\Delta\lambda^2 + \frac{\Delta\lambda^3}{3} \right) \tag{VII-2}$$

If $\Delta\lambda = 1$ and $\lambda \sim 10^4$, the last two terms within the parentheses may be neglected. Hence,

$$V \simeq 2\pi\lambda^2 \tag{VII-3}$$

The number of collisions η per unit volume during time ΔT is then given by

$$\eta = \frac{\Delta N}{V} = \frac{N_oS_ov_o\Delta T}{2\pi\lambda^2} \tag{VII-4}$$

Not all collisions lead to ionization, however. If α is the coefficient of ionization, the number of ionization events is given by

$$\eta^* = \alpha\eta = \alpha\frac{N_oS_ov_o\Delta T}{2\pi\lambda^2} \tag{VII-5}$$

At sufficiently large λ, the satellite will collide only with an insignificant part of the ions it produces. Also, the ions that are produced only have thermal velocities. It can be theoretically shown that the ions that can be incident on the satellite come only from that portion of the spherical surface which is bounded by a circle of radius $\lambda/20$. Thus, the number N^* of the newly formed ions on this sector of the surface is given by

$$N^* = \eta^*\pi\left(\frac{\lambda}{20} \right)^2 \tag{VII-6}$$

where

$$\eta^* = \frac{\alpha_oN_oS_ov_o\Delta T}{8 \times 10^2}$$

Since only a small fraction of N^* falls on the satellite surface, the net number of ions per unit volume is

$$\eta_1^* \simeq \frac{5\alpha N_oS_ov_o\Delta T}{32\pi\lambda^2} \tag{VII-7}$$

Assuming $\alpha \sim 10^{-4}$ (according to Gerlofson), $N_0 \sim 10^{10}$ at an altitude of 250 km, $\lambda \sim 10^4$ cm, $S_0 \sim 4 \times 10^4$ cm^2, $v_0 \sim 8 \times 10^5$ cm/sec, and $\Delta T \sim \frac{1}{8} \times 10^{-5}$ sec. Equation (VII-7) yields $\eta_1^* \approx 20$ ions/cm^3.

Thus the artificial ionization of 20 ions/cm^3 resulting near the surface of the satellite and induced by its motion is negligible compared to the natural ionization of the order 10^5 to 10^6 ions/cm^3. The majority of the ions produced have no chance at all to return to the satellite. It can therefore be assumed that the satellite moves in an undisturbed medium.

Another important phenomenon associated with the high velocity of an artificial satellite is the deep vacuum created in the wake. This vacuum is caused by its high velocity relative to the medium. Measurement of the ion composition of the atmosphere within the vacuum cone will be a complete failure because an instrument placed in the "rear cone" will be totally unable to operate due to insufficient ion density. For reliable data, it is absolutely essential that the satellite is properly oriented or stabilized.

3. INSTRUMENTS FOR THE DIRECT STUDY OF THE ION AND NEUTRAL COMPOSITION OF THE UPPER ATMOSPHERE

While the gas composition of the atmosphere can be analyzed by several different methods, a study of the ion composition of the atmosphere is only possible with the use of rocket- and satellite-borne mass spectrometers. Like any other instrument on a space vehicle, the mass spectrometer must operate automatically and must have a minimum time lag. It should meet the requirements of mechanical strength, endurance of large overloads, endurance of vibration, large temperature resistance, good vacuum conditions, etc. The cumbersome instrumentation of magnetic mass spectrometers inhibits the use of these instruments in ionospheric research. The huge and heavy magnetic instruments, in spite of their high resolution and high sensitivity, are outmoded in this era of miniaturization. The nonmagnetic mass spectrometers are designed to meet the requirements of miniaturization and they are sensitive enough to faithfully report to a considerable degree of accuracy the neutral and ion composition of the upper atmosphere.

The nonmagnetic instruments that are used are the Bennett-type radio-frequency mass spectrometers, the time-of-flight mass spectrometers, and the quadrupole mass spectrometers. Recently Narcisi *et al.* have used the quadrupole mass spectrometer quite successfully to investigate the ionic content of the D region.

The RF mass spectrometer can be used for the analysis of either neutral or ionized gases contained in the earth's atmosphere. When the instrument is used for the investigation of the ion content of the atmosphere, there is no need for an ion source since the ionosphere itself is one. However, the instrument needs an ionizing source when it is used to measure the neutral gas composition. The mass interval in which the instrument operates is determined by the geometric dimensions of the analyzer, the operating frequency, and the range of variation of the accelerating voltage. The altitude interval in which the instrument operates is determined by (1) the geometric dimensions of the tube (mean free path > length of RF analyzer), and, (2) density of ions entering the analyzer. When the positive ion concentration is of the order of 10^5 cm^{-3}, the collector ion current may reach 10^{-8} A. The instrument apparently can operate for still lower concentrations with suitable alterations in the design.

4. COMPOSITION STUDIES USING ROCKET- AND SATELLITE-BORNE MASS SPECTROMETERS

Johnson, Meadows, Townsend, *et al.* initiated the mass-spectrometric investigation of the ion and neutral gas composition of the upper atmosphere by launching rocket-borne RF mass spectrometers in 1954 and 1955. Since 1955, many people have been flying mass spectrometers. Everyone is confronted with some sort of experimental difficulty, and the results obtained by each of these investigators are extremely controversial,* but challenging. The conclusions drawn by the earlier investigators have had to undergo such radical changes that when we look back at their speculations today, they read rather strangely. However, one must commend the patience and enthusiasm of the early workers. Upper atmospheric physics is in a great state of confusion, theories coming and going. If the speculations on the atmospheric composition based on available mass spectra are not reliable, it can only be attributed to insufficient data. Only recently have we been obtaining more and more data, so that an investigator is able to take a stand with confidence when he ventures to make any speculations. It is only

---

* A particular mass number could signify different types of molecules. If one speculates about the constituents based on the mass spectra one obtained, one is always guided by: (1) the available knowledge of the physical processes which justify or rather necessitate the existence of that constituent, and (2) physical processes which could only be explained by the presence of the constituent speculated.

attempted here to review and report the various experiments that were conducted since the advent of mass spectrometry into upper atmospheric research. Both successes and failures, false and true speculations are reported. The main concern of the author is to trace the history of mass-spectrometric research during its early period.

a. Townsend's Experiment

Townsend (1952) developed the electronic circuits for the Bennett-type RF spectrometer so that it could be used in an experiment devised to obtain the composition of the upper atmosphere. The rocket-borne spectrometer was able to sample the spectrum between mass numbers 5 and 48 once per second with a resolution of 1:25. Data could be obtained within pressure limits of 8×10^{-4} and 2×10^{-6} mm Hg corresponding to the ambient pressures at 100 and 160 km, respectively. The entire instrument weighed less than 50 lb and occupied a total volume of 1.4 ft³. The mass spectrometer tube used by the investigators is a slightly modified version of the Bennett 7–5-cycle RF spectrometer. It is shown in Fig. VII-1. The associated circuitry is shown in Figs. VII-2 and VII-3.

It was necessary to modify the Bennett tube by the addition of a final grid held at a potential of -300 V. This grid suppresses the negative background at the collector caused by secondary electrons arising at the tungsten grid and metal surfaces in this region. The spectrometer tube was shock-mounted in a bed of foam rubber and placed within a pressure-tight aluminum box. In case of a failure in the glass envelope of the tube, this box could prevent evacuation of the rocket instrument section which was pressurized at one atmosphere. All leads were brought out at the rear of the tube through a standard 14-pin cathode-ray tube base and by-passed to ground through ceramic capacitors. Only the collector lead was shielded with a brass tube and terminated in a coaxial fitting. The focus condition applicable to this instrument is (in amu)

$$M = \frac{0.266V}{S^2 f^2} \tag{VII-8}$$

where f is the frequency in megacycles, V is in volts, and S is in centimeters.

The tube was swept over a frequency range corresponding to the masses of common atmospheric gases and the relative abundances of the various constituents could be obtained. Only the frequency sweep was utilized since it was difficult to design a simple

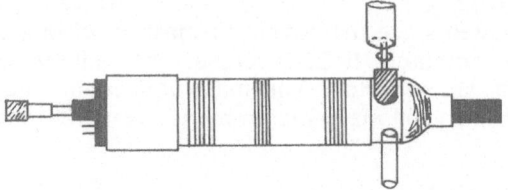

After J. W. Townsend, Jr., *Rev. Sci. Instr.* **23**, 538, (1952).

Fig. VII-1. Townsend's radio-frequency mass spectrometer tube.

sweep oscillator with a high degree of amplitude stability. The sweep rate was limited to a maximum value of 1/sec by the frequency response of the telemetering system in use. This rate was adapted for use in the rocket, since the pressure would fall rapidly around the ascending missile, making the highest possible sweep rate mandatory. With a frequency of 3.9 Mc and a sweep varying between -250 V and -25 V, the instrument could cover the range between 48 amu and 5 amu once per second.

While analyzing the circuit, we notice that the RF voltage for the analyzer is supplied by an oscillator of the modified Pierce type, the frequency being controlled by a 3.9-Mc crystal in a feedback loop. (Fig. VII-3). The output voltage of the oscillator is varied by a potentiometer in the screen-grid circuit and is adjustable to provide a range of 3 V rms to 9 V rms at the spectrometer tube. The output of the oscillator is also fed to two voltage-doubler rectifiers via decoupling resistors to obtain the bias and stopping potentials.

The one-cycle time-base generator employs a 6AS6 Miller Integrator with an infinite external plate load. An 884 Thyratron starts each cycle after the rundown period. With an average current drain of less than 1.0 mA, a 250-V sweep is avoidable with an excellent linearity and short fly-back time. A small 28-V dynamotor supplies the 300 V required for the oscillator and sweep circuit.

A 100% feedback (regular) DC amplifier was chosen to detect the ion current. An important consequence of the feedback in this type of amplifier is the reduction of the input time constant to an apparent value given by

$$T = \frac{R_g C}{1 + G} \qquad\qquad \text{(VII-9)}$$

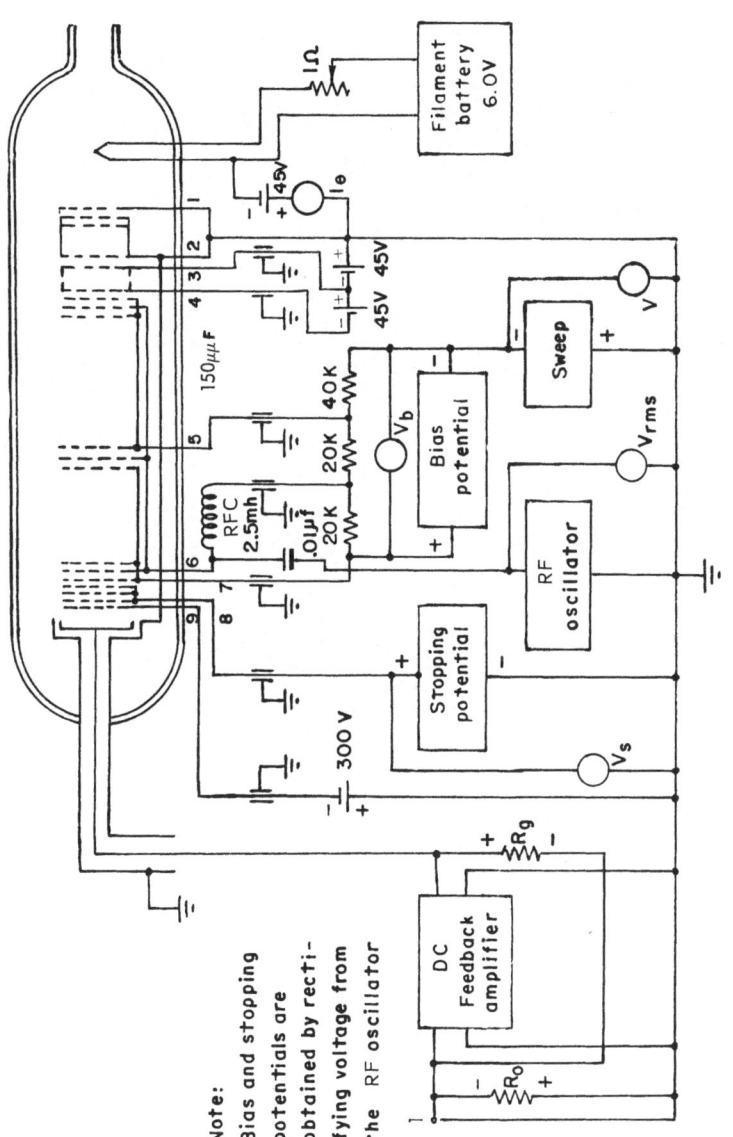

Fig. VII-2. Townsend's mass spectrometer block diagram.

After J. W. Townsend, Jr., *Rev. Sci. Instr.* 23, 539, (1952).

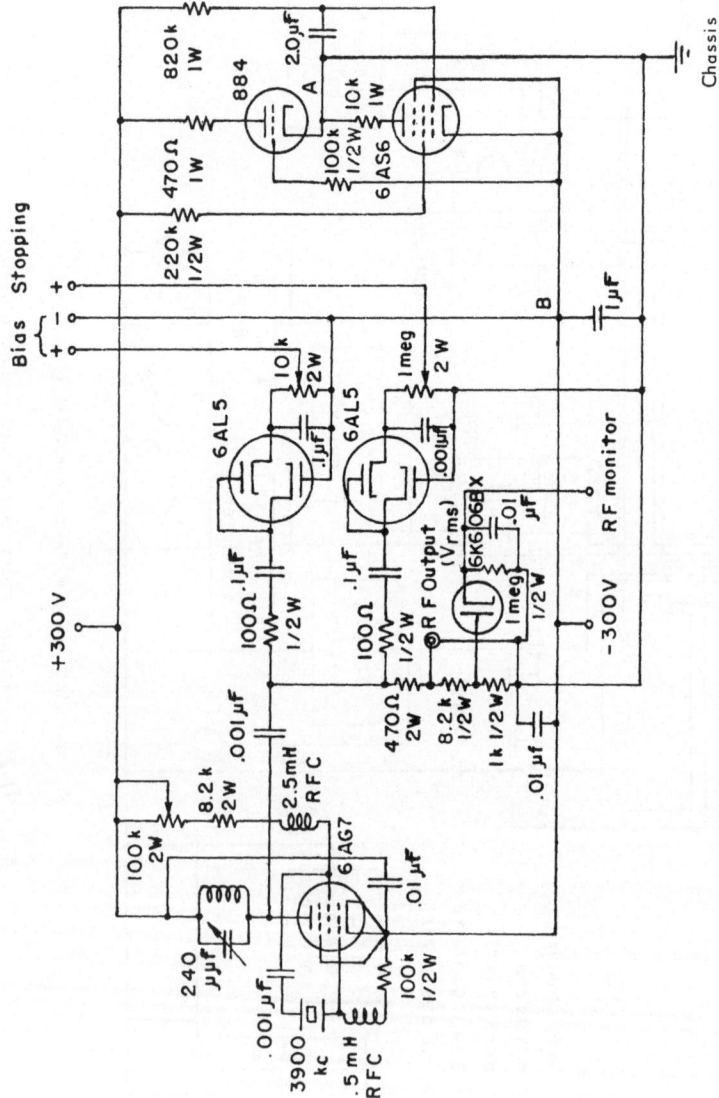

After J. W. Townsend, Jr., *Rev. Sci. Instr.* **23**, 539, (1952).

Fig. VII-3. Townsend's mass spectrometer electronics circuitry.

where R_g is the input grid resistor and C is the total collector-tube capacity to ground. Then the output voltage is given by

$$V_o = iR_g \frac{G}{G + 1} \qquad \text{(VII-10)}$$

where i is the input ion current and G is the amplifier gain. Thus the output voltage is proportional to the input voltage. Figure VII-4 represents the detector that was used in the experiment.

The amplifier of the detector saturates at an input signal of 30 V. For the rocket installation, R_g was set at $5 \times 10^8 \, \Omega$ so that the molecular nitrogen peak just begins to saturate at the highest operating pressure (8×10^{-4} mm Hg). The output voltage is then taken across the 20 K Ω potentiometer R_o and divided so that two gains might be fed to the 0 to 5V DC telemetry channels (points 2 and 3 in Fig. VII-4). The detector tube was separately packed.

The spectrometer was packaged into four units: the tube in its pressure-tight box, the electronics chassis, the detector chassis, and the battery box. The battery supplying the spectrometer filament (3.8 A) limited the operation of the experiment to about 15 min, which is more than enough. The installation of the apparatus was made at the base of the nose cone on the Navy *Viking* rocket. The entire instrumentation was well shielded and employed a common ground point to avoid noise pickup and interference from other rocket functions.

Before flight, the tube was thoroughly outgassed and repressurized with one atmosphere of helium. The rubber stopper, initially waxed in place, was forced out by the pressure differential shortly after take-off, thus opening the inlet tube. At 100 km (pressure of 8×10^{-4} mm Hg), a timer would automatically turn on the spectrometer filament to begin operation. Calibration with a known mixture of gases was done in the laboratory over the operating range of pressures. The need to keep the filament of the spectrometer tube well away from the ionization region in order to prevent absorption was also recognized. The reproduction of a typical record of air taken in the laboratory is shown in Fig. VII-5.

b. Experiment of Johnson and Meadows

Johnson and Meadows (1955) made the first investigation of ambient positive ion composition from 93 to 219 km using a rocket-borne Bennett-type RF mass spectrometer, perfected for rocket use by Townsend. The experiment was flown in the Navy *Viking 10* rocket at White Sands, New Mexico. The results of this experiment are not to be trusted, since experimental complications bred so much unreliability into the data that the results can only

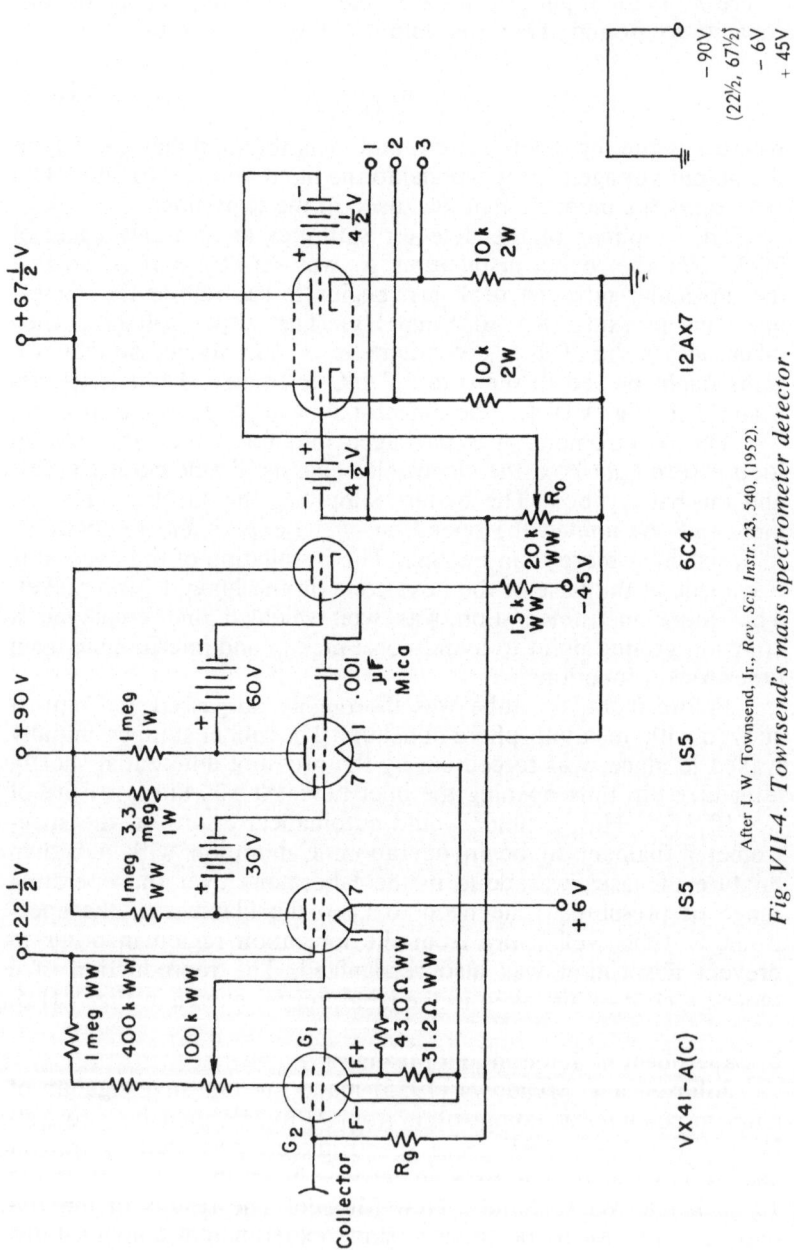

After J. W. Townsend, Jr., *Rev. Sci. Instr.* **23**, 540, (1952).

Fig. VII-4. Townsend's mass spectrometer detector.

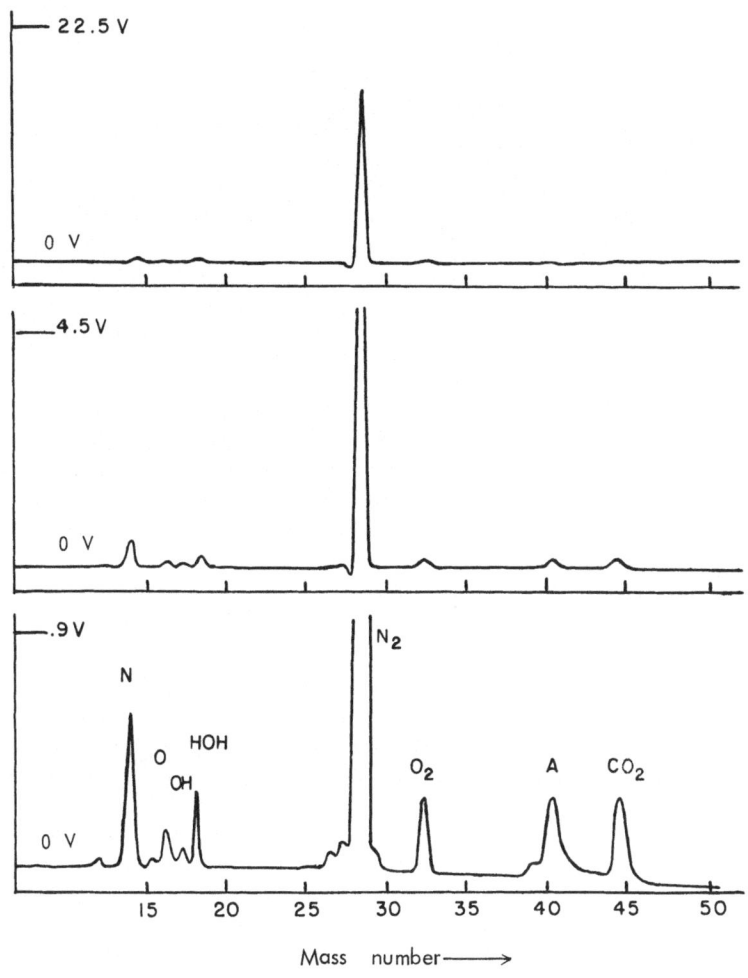

After J. W. Townsend, Jr., *Rev. Sci. Instr.* **23**, 541, (1952).

Fig. VII-5. Townsend's typical laboratory spectra of air.

be treated as preliminary. Also, at that time, the vehicle (rocket) contamination and the potential acquired by the rocket were two major sources of error which had not yet been successfully tackled.

The tube was attached to the rocket skin 2.1 m from the nose tip and perpendicular to the rocket axis, by means of an adapter ring with two O-ring seals to maintain nose-cone pressurization. The 4.1-cm-diameter hole in the rocket skin had a pumping speed

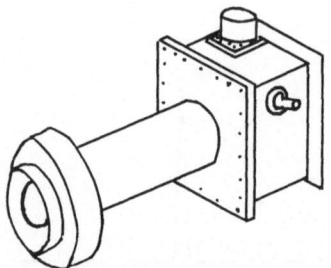

After C. Y. Johnson and E. B. Meadows, *J. Geophys. Res.* **60**, 194, (1955).

Fig. VII-6. Johnson and Meadows' radio frequency mass spectrometer for Viking 10.

of 150 liters/sec. The flight unit, shown in Fig. VII-6, was 25.4 cm long overall and weighed 2.2 kg. Figure VII-7 shows the instrumentation of the positive-ion spectrometer for the *Viking 10* rocket. Positive ions were drawn into the spectrometer by the negative potential on the first grid and further directed down the tube by the potential on the screen grid.

The mass number condition is

$$M = \frac{0.266(V_S + kV_{rp} + 2.05\ V_{rms})}{S^2 f^2} \tag{VII-11}$$

where V_{rp} is the apparent rocket potential, V_{rms} is the rms value of the RF voltage, and k is a fractional constant. A negative charge on the rocket would result in a mass scale which is shifted toward lower mass numbers.

Adjustment of the spectrometer prior to flight was based on the expectation that the rocket would acquire a negative charge. Since the rocket potential due to this negative charge effectively lowers the stopping potential, the stopping potential was set close to that required for complete suppression of ion current. This left an operating voltage of 10 V, before harmonics would appear. The absolute mass scale for the flight spectra was determined by measuring the sweep potential from the telemetered data, and the mass scale extended from 8.2 to 49.2 amu. At 93 km, there appeared a conspicuous ion-current peak at mass 32. At 106 and 107.5 km, the positive-ion peaks at masses 12, 16, 21, 23, 26, 30, 32, and 48(?) were detected. The behavior of the various peaks with altitude on the ascent was also studied. Peaks at mass numbers 16, 23, 32, and 48(?) persisted up to 219 km. Above 120 km, it was expected that harmonic peaks would appear due to the rocket potential. The lower harmonic for the peak at mass 32 is at mass 23.8,

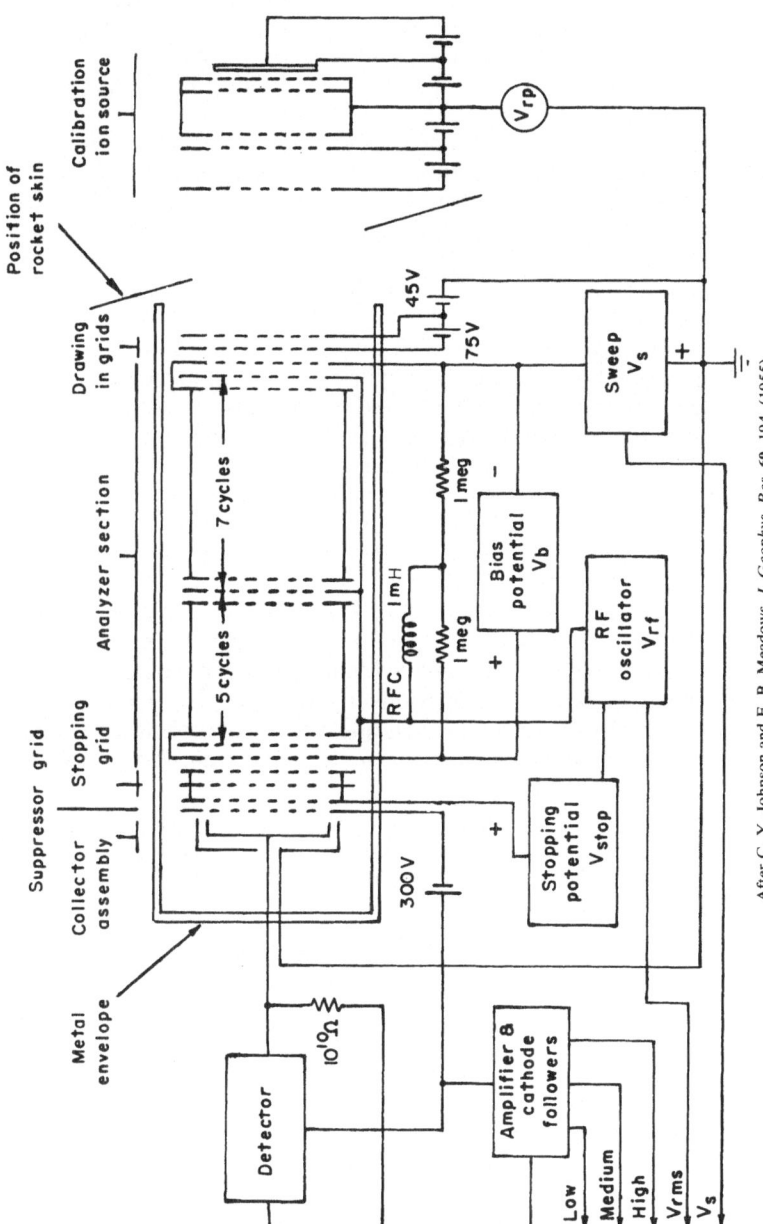

After C. Y. Johnson and E. B. Meadows, *J. Geophys. Res.* **60**, 194, (1955).

Fig. VII-7. Johnson and Meadows' diagram of the positive-ion spectrometer instrumentation for Viking 10.

and for mass 48 is at mass 35.6. But a peak was observed only at 23.8. Since no peak was observed at 35.6, it is extremely doubtful whether the peak at 23.6 is actually the lower harmonic of 32.

The clarity of the flight spectra was impaired by the presence of oscillations and noise, and by the broadness of the peaks. The ion peaks were very broad due to the shift experienced from the rocket potential. The distortion in the spectra at lower altitudes can be explained in the following fashion. The rocket was on its side and was spinning so that the mean free path in the spectrometer was undergoing a continuous change. When the mean free path in such a spectrometer reduces to 30 cm or less, the efficiency of the spectrometer is automatically reduced. This is why the observed spectra were extremely dependent on the spatial orientation.

The cause for the rocket acquiring a negative potential has been speculated by many people and many theories are being presented to explain the negative charge. It may be apparently due to electrons striking the rocket with energies greater than 20 eV. The electrons may realize this energy from the RF field of any high-frequency transmitter antenna aboard the vehicle, or by photo-electric absorption of the sun's high-energy X-rays by the gas around the rocket.

The investigators (Johnson and Meadows) at that time tagged the various observed mass numbers with different atoms and molecules. Some of these mass number identifications have been shown to be wrong in later investigations by other experimenters (discussed later). However, an earnest quest toward forming an atmospheric model by direct investigation was initiated. The specula-

Table VII-1.   Speculations by Johnson and Meadows on Mass Number Identifications

Mass Number	Chemical Composition
12	Carbon (C)
16	Oxygen (O), atomic
18	Water vapor (H_2O)
19	Fluorine (F) or (H_3O)
21	?
23	Sodium (Na)
26	Cyanogen (CN), Acetylene (C_2H_2)
30	Nitric oxide (NO)
32	Molecular oxygen (O_2)
38	?
45	?
48 (?)	?

tions of Johnson and Meadows are presented in Table VII-1. Figure VII-8 represents the positive-ion spectra at different altitudes taken by an RF mass spectrometer aboard *Viking 10*. Figure VII-9 represents the ion currents of various constituents, and rocket potential versus altitude.

c. Experiment of Johnson and Heppner

The nighttime ambient positive- and negative-ion composition in the mass range 5 to 58 amu was measured by Johnson and Heppner (1955) between 98 and 120 km by two Bennett RF mass spectrometers on an Aerobee rocket, NRL No. 23, at New Mexico, at 01:39 MST on July 8, 1955. The rocket did not acquire measurable potential nor was a large quantity of rocket gas (contamination) observed, as was the case in the positive-ion spectrometer flown before in *Viking 10*. Positive ions of mass number 28 were observed and ions of mass numbers other than 28 would have been detected even if their concentration had been as low as 1% of the concentration of mass 28 ions. Mass 28 was attributed to N_2^+ and the possibility that it could be CO^+ seems remote, considering the very low relative abundance of neutral CO molecules. There was no indication of negative ions at the operating altitudes of the spectrometer. It was claimed that negative ions would have been detected if their concentration had been as low as 1% of concentration of N_2^+ ions. However, there remained unanswered many intricate problems concerning the experimental technique of detection of negative ions.

Also, a contradictory result was observed by the investigators. Ionospheric records during the time the rocket was in the E region showed sporadic E reflections at maximum frequencies varying between 2.8 and 5.0 Mcps at a virtual height of 100 km. The various theories set forth to explain nocturnal E region ionization at the time the investigators conducted their experiment had quite generally attributed this ionization to oxygen atoms and molecules and/or oxides such as NO. It is thus particularly significant that in this measurement the dominant ion was found to be N_2^+.

d. Experiment of Townsend and Meadows

In 1956, Townsend and Meadows reported another investigation of the neutral gas composition of the upper atmosphere during a nighttime flight using the same Bennett-type RF mass spectrometer. Increased resolution was obtained by using a four-stage tube in this experiment. From 113 km to 141.6 km (the peak altitude) on the ascent of the rocket, and from 141.6 km to 74 km on the descent, the spectrometer detected the usual constituents

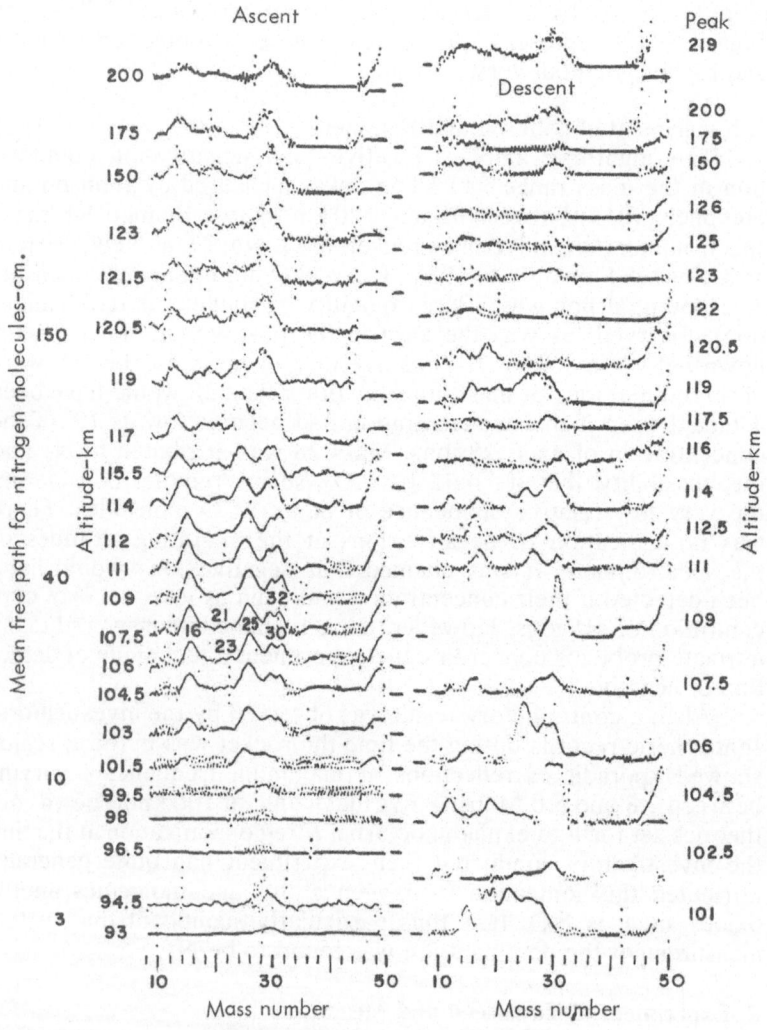

After C. Y. Johnson and E. B. Meadows, *J. Geophys. Res.* **60**, 198, (1955).

Fig. VII-8. Johnson and Meadows' positive-ion spectra: $10^h/00^m$, May 7, 1954, White Sands Proving Ground, New Mexico.

of air. In the vicinity of 85 km on the descent, components of mass numbers 46 and 23 were detected and were tentatively identified as nitrogen dioxide (NO_2) and sodium, respectively.

The peaks of mass numbers 23 and 46 were detected between 88 and 80 km (on descent) and between 75 and 90 km (on ascent). These peaks were tentatively identified as sodium and nitrogen dioxide, respectively. In the photometer experiment flown by Koomen, Scolnik, and Tousey in the same rocket, sodium D radiation was found to be the most intense at about 85 km, with the top and bottom of the luminous layer at 100 and 80 km.

5. IGY ROCKET MEASUREMENTS OF ARCTIC ATMOSPHERE COMPOSITION ABOVE 100 km

a. Composition Studies of Meadows and Townsend

Five Aerobee-Hi rockets containing RF mass spectrometers for the study of gaseous and ionic composition in the E and lower F regions of the atmosphere (above 90 km) were launched from 1956 to 1958. Four were launched at Fort Churchill, Canada, and the other at White Sands, New Mexico. The first four were launched in the arctic region. Table VII-2 gives the launching details about the rockets. Spectrometers covering the approximate mass range of 10 to 50 amu were utilized. The spectrometer used for gas analysis was of four-stage design and the one used for positive- and negative-ion analysis was a three-stage tube. Figure VII-10 is a photograph of the type of instrument used.

The ion spectrometers were mounted in an extension to the rocket body located just below the nose cone assembly. The ion spectrometer openings were on opposite sides of the rocket, and the spectrometer axes were normal to the rocket's main axis. The spectra were taken once every second and the resolution of the

Table VII-2. Details of Launching of Four Aerobee-Hi Rockets at Fort Churchill, Canada

Rocket	Date	Time (CST)	Peak altitude	Weather conditions
IGY NN 3.17 NRL-48	Nov. 20, 1956	23:21	251 km	Overcast
IGY NN 3.18 F	Feb. 21, 1958	20:02	225 km	Aurora
IGY NN 3.19 F	Mar. 23, 1958	12:07	203 km	Polar blackout
IGY NN 3.20 F hydrogen – helium experiment	Nov. 3, 1958	10:20	209 km	

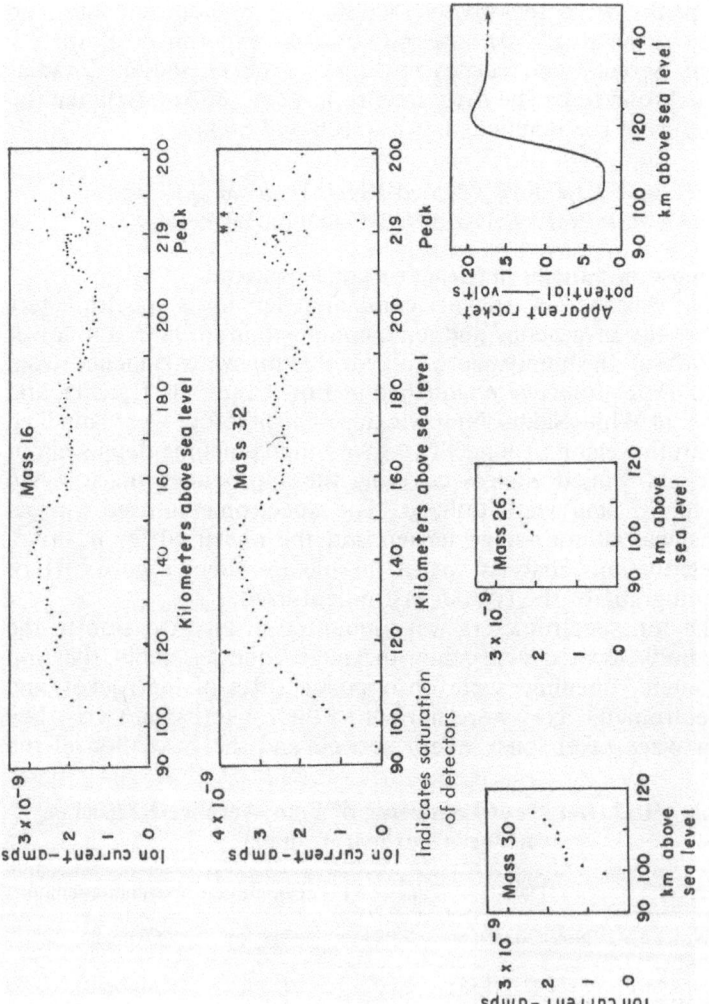

After C. Y. Johnson and E. B. Meadows, *J. Geophys. Res.* **60**, 200, (1955).

Fig. VII-9. Johnson and Meadows' ion current of ions of 16, 26, 30, and 32 amu, and apparent rocket potential vs. altitude.

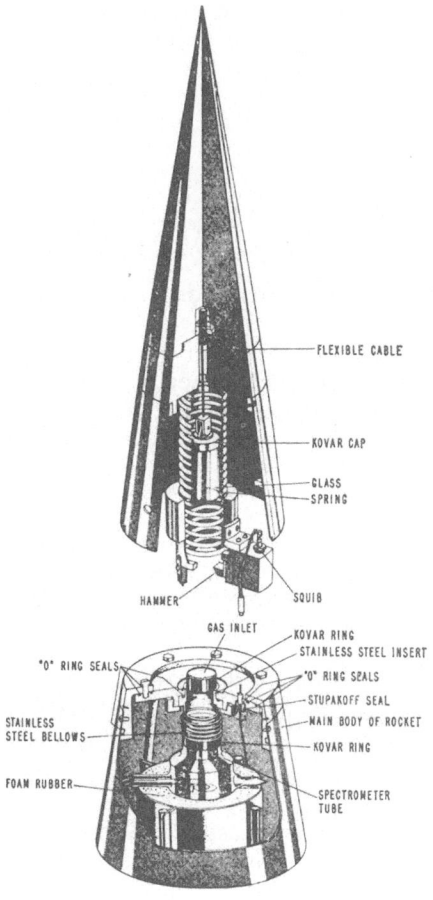

FLEXIBLE CABLE

KOVAR CAP

GLASS
SPRING

SQUIB

HAMMER

GAS INLET

KOVAR RING
STAINLESS STEEL INSERT

"O" RING SEALS

"O" RING SEALS

STUPAKOFF SEAL

STAINLESS
STEEL BELLOWS

MAIN BODY OF ROCKET

KOVAR RING

FOAM RUBBER

SPECTROMETER
TUBE

After E. B. Meadows and J. W. Townsend, Jr., from p. 177, *Space Research—Vol I* edited by H. Kallmann-Bijl, North-Holland Publishing Co., Amsterdam (1960).

Fig. VII-10. Rocket flight instrument used by Meadows and Townsend.

four-stage gas spectrometer used was about 40. The spectrometer could operate when the mean free path was equal to the length of the analyzer (pressure $< 1\,\mu$), and was calibrated in the laboratory by admitting air into the spectrometer through a ceramic leak. A calibration was also performed for various gas pressures in the tube in the pressure range 10^{-3} to 10^{-5} mm, since below this range, the sensitivity of the instrument is independent of pressure.

Some physical problems could be discussed on the basis of the composition data obtained by these rocket flights into the arctic sky.

i. Diffusive Separation. The diffusive separation ratio r at an altitude h is defined by

$$r = \frac{(I_{argon}/I_{nitrogen})_h}{(I_{argon}/I_{nitrogen})_{ground\ level}} \qquad \text{(VII-12)}$$

where I stands for the current due to the particular constituent. If $r = 1$, no separation is present. If $r = 0$, separation is complete. Argon and nitrogen are chosen in the definition because (1) argon is inert and is not expected to occur as an atmospheric ion, and (2) nitrogen is relatively nonreactive and is not expected to occur dissociated or ionized appreciably. The separation ratio is given by

$$r = e^{-\Delta h/H_1} \qquad \text{(VII-13)}$$

where H_1, the relative scale height $= RT/g(M_{Ar} - M_{N_2})$, Δh is the arbitrary altitude interval in question, R is the universal gas constant, T is the average temperature in degrees Kelvin, M_{Ar} is the mass number of Argon (40), and M_{N_2} is the mass number of molecular nitrogen (28).

Townsend *et al.* thought that diffusive separation might be possible below 100 km. However, the time required is too long and turbulent mixing forces are too strong. Above 120 km, the time required for diffusive separation is quite short and the investigators expected a transition region between 100 km and 120 km, where the time required to establish diffusive equilibrium is approximately equal to the lifetime of a disturbance resulting in turbulent mixing. It should also be noted in passing that the distribution of hydrogen is not expected to be governed primarily by diffusion in this altitude region.

ii. Mean Molecular Weight. Figure VII-11 gives the mean molecular weight as a function of altitude computed directly from the recorded spectra. Another valid observation that has been made by the investigators is that an extrapolation of the experimental data down to 100 km gives an average molecular weight near the ground value. This agrees with the observation of turbulent mixing below this altitude.

Observation of atomic and molecular oxygen densities led the investigators to believe that the measured scale heights must be

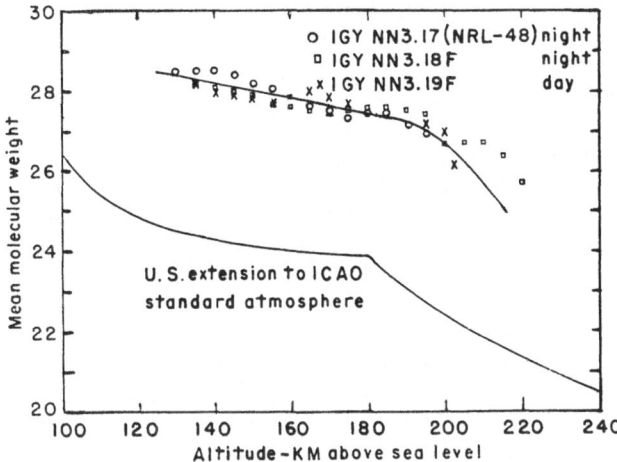

After E. B. Meadows and J. W. Townsend. Jr.. from p. 184. *Space Research—Vol. I* edited by H. Kallmann-Bijl. North-Holland Publishing Co.. Amsterdam (1960).

Fig. VII-11. Mean molecular weight of atmospheric gases in the arctic atmosphere.

interpreted in terms of a sharp increase in atmospheric temperature over Fort Churchill between altitudes of 120 to 180 km. This temperature increase could affect the diffusion process and thereby decrease the atomic to molecular oxygen ratio in this region. There is evidence for this high temperature gradient in the thermosphere from other sources. However, the densities observed at high altitudes by satellites were greater than expected and Nicolet has interpreted this in terms of the Chapman theory of atmospheric heating due to heat conduction from the extension of the solar corona in the vicinity of the earth. Nicolet concluded that there must be very high temperatures in the thermosphere. Another possible cause suggested for high temperatures in the arctic thermosphere could be the direct heating due to absorption of energetic particles trapped in the Van Allen belt near the geomagnetic poles. Fort Churchill is located at the end of the belt in the auroral zone.

iii. Atmospheric Density. Using the mass spectrometer as a pressure gauge, the investigators derived a density distribution in the lower thermosphere. For this purpose, equations of Horowitz and LaGow were adapted. However, the equations used are by no means general. Ambient density is given by

$$\rho_a = \frac{P_g}{V \cos \sigma (2\pi R' T_g)^{1/2}} \qquad (VII\text{-}14)$$

where P_g is the spectrometer pressure, V is the rocket speed, σ is the angle between the rocket axis and relative wind direction, $R' = k/m$ where k is the Boltzmann constant, and T_g is the absolute temperature of the spectrometer tube. Figure VII-12 shows the ambient density at various altitudes.

iv. Positive-Ion Composition. On the basis of the data obtained on individual detected constituents, little could be said concerning the remaining gases. Masses 44, 30, and 18 were attributed to CO_2, NO, and H_2O, respectively. These gases were not present in appreciable quantities in the spectra. Townsend *et al.* conclude that the CO_2 and H_2O found during the flight were not necessarily of ambient origin, although in part they might be. The discovery of NO in the spectra led people to speculate about its formation as a result of charge exchange in the ion source.

Normally the concentration of NO molecules is inferred from the intensity of the solar Lyman-α line and the electron density of

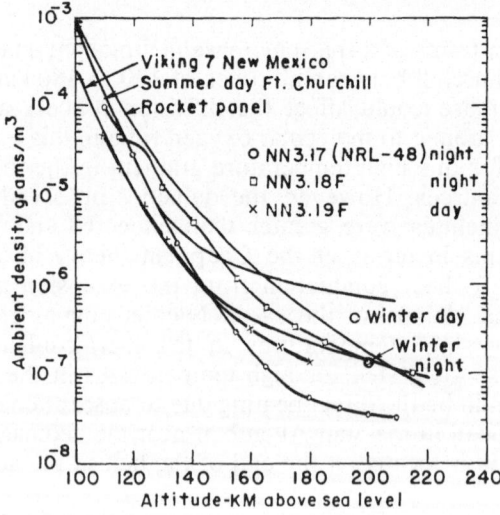

After E. B. Meadows and J. W. Townsend, Jr., from p. 187, *Space Research—Vol. 1* edited by H. Kallmann-Bijl, North-Holland Publishing Co., Amsterdam (1960).

Fig. VII-12. Atmospheric density at various altitudes.

the D region. It is generally believed that the D region is formed as a result of photoionization of NO under irradiation by Lyman-α. It can be shown that less than one part per million of NO is sufficient to produce the observed D region electron density. Because the ionization potential of NO is only 9.4 eV, much less than the ionization potentials of the common atmospheric molecules, nitric oxide can charge-exchange with any of the other atmospheric ions. Nicolet has suggested that the reaction

$$N_2^+ + O \rightarrow NO^+ + N \qquad \text{(VII-15)}$$

has a cross section high enough to account for the high abundance of NO^+.

v. General Results. Also, it was felt necessary to provide cleaner surfaces in and around the spectrometers to reduce the hydrogen background. The helium background could be made negligible by pressurizing the propellant tanks with nitrogen instead of helium.

From the rocket flights into the arctic skies, the following conclusions were reached: (1) Diffusive separation occurs in arctic upper atmosphere both day and night in the late fall, winter, and early spring seasons. The maximum level is about 100 to 120 km. (2) Influence of winds and turbulent disturbances in these regions seemed to merit serious investigation. (3) Based on available data on atmospheric parameters, no decision could be taken as to the transition region of diffusive separation. (4) Mean molecular weight of the atmospheric gases in the arctic thermosphere appeared to be probably higher than previously believed. The situation at lower altitudes could be different.

b. Positive-Ion Studies of Johnson and Holmes

At the same time when Townsend *et al.* measured neutral and ion gas composition of the arctic atmosphere with Aerobee-Hi rockets during the International Geophysical Year, Johnson and Holmes measured positive-ion content with the help of ion mass spectrometers on the same flights. Three of the rockets carried two mechanically identical mass spectrometers of the Bennett type: one for positive-ion analysis in the range 9 to 50 amu, and the other for negative ions in the same mass range. The fourth rocket carried two positive-ion spectrometers, the first one covering the range 0.5 to 5 amu and the second one 9 to 35 amu. In addition, each rocket carried a conventional Bennett mass spectrometer to examine the neutral atmospheric constituents.

Each ion spectrometer was encased in a tube, the open end of which was fastened to the rocket skin. Ambient ions enter through this opening in the side of the rocket and are analyzed according to mass by the Bennett system of grids. For a fixed rocket potential, there is a one to one correspondence between an ion mass and the instantaneous linear sweep voltage V_A, which assigns to each mass a fixed position in the time sweep. There is no internal ion source in this case as the ambient ions of the ionosphere form a fresh source and thus solve many experimental problems. There is no pumping required to assure that the supply of ions is "fresh." The relative ion currents were corrected for vehicle attitude, vehicle potential, and vehicle velocity. However, the instrument, as designed, could not be operated below 85 km.

Some of the results obtained seem rather interesting and the speculations made by the investigators as to the chemical nature of the detected ionic masses are partly controversial and partly erratic. But anyway, as pointed out earlier, nothing more could be done by the researchers at that stage.

Mass 30^+ was the predominant ion detected in the E region and mass 16^+ seemed abundant in the F region. Mass 32^+ was detected during a polar blackout experiment and this was related to production of 32^+ by hard X-rays in the auroral zone. Since no positive ions in the mass range 0.5 to 5 amu were detected, people began to believe that the amount of positively ionized hydrogen is less than 10^3/cc below 230 km.

It seemed that the three ionic species, 16^+, 30^+, and 32^+, accounted for more than 95% of the total ion current at any altitude.

One of the rockets, deliberately aimed toward an aurora, detected mass 28^+ in the 100 km region. First, it was thought to be of auroral origin, but it was later suggested that this ion is a contaminant generated by the neutral gas mass spectrometer which was also flown with the ion mass spectrometer. The investigators were not in a position to discuss the origin of the ions of masses 28^+, 14^+, and 18^+ which appeared as minor constituents.

6. AN EMISSION CURRENT REGULATOR FOR RF MASS SPECTROMETERS

In 1957, Holmes developed a new design for an emission current regulator for rocket-borne Bennett-type RF mass spectrometers. The maintenance of constant emission currents in the laboratory mass spectrometers is not a problem. However, the use of rockets for research vehicles has created the need for new requirements on the weight, size, and power consumption of all electronic instrumentation. Certain rocket-borne RF mass spectrometers

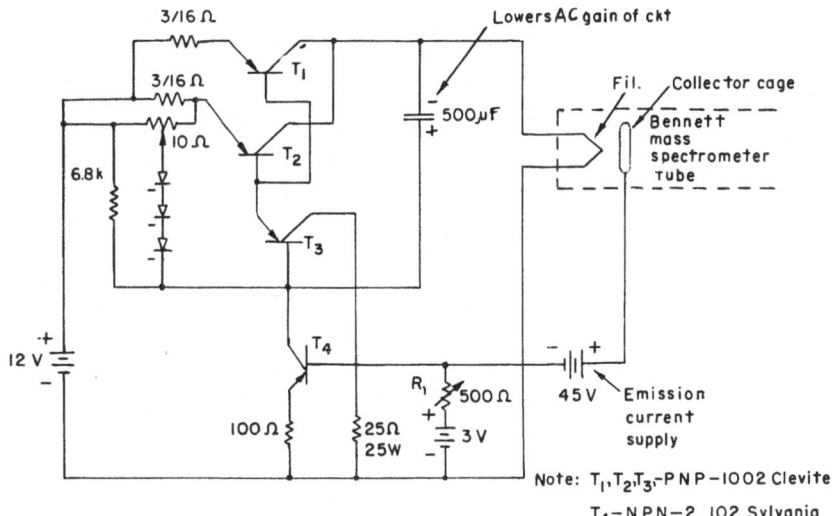

After J. C. Holmes, *Rev. Sci. Instr.* **28**, 290, (1957).

Fig. VII-13. Diagram of Holmes' emission-current regulator.

used to have rheostats in the filament lead which are preset to produce the desired emission current. The weight and power consumption of conventional emission regulators had precluded the utilization of such devices in rockets. The advent of transistors did change this situation, and a transistorized series-type regulator was developed by Holmes which could handle directly the 3- to 5-A currents required by the mass spectrometer filament. Figure VII-13 shows the diagram of the emission regulator.

Under pressure changes inside the mass spectrometer of from 10^{-4} mm to 10^{-6} mm Hg, the emission current variation is less than 1% when using this regulator. Filament battery-supply voltage changes of 20% produce less than 2% change in the emission current. The complete regulator unit including heat-sink for the power transistors weighs less than 12 oz.

7. A THEORY TO DETERMINE IONIC MASSES IN RF SPECTROMETERS IN FLIGHT

Johnson in 1958 advanced a theory toward the determination of mass of ions detected by the Bennett-type RF ion mass spectrometer. The use of the Bennett RF spectrometer as an upper-atmospheric research tool is based on its light weight, high current efficiency, and large sampling area. In ion mass spectrometry, the ion source used for calibration purposes in the laboratory is detached from the spectrometer prior to the flight and the ionosphere becomes the ion source during the flight. Since the mass scale of a Bennett spectrometer is dependent on the ion entrance energy, this change of ion source requires a method whereby the absolute mass of ions detected during flight can be determined. The following method due to Johnson provides this absolute in-flight mass determination. It makes use of the "harmonics" present in the mass spectra obtained. Application of this method to some ion composition data resulted in a clarification of the results.

If v is the velocity of the ion through the spectrometer and is a constant, then the energy gain produced by the RF voltage in a multistage tube (see Wherry and Karasek) is

$$\Delta W = (-2V_{RF}e)\left[1 - \frac{\cos(S\omega/v)}{S\omega/v}\right]\left\{\cos\left(\frac{S\omega}{v} + \theta\right)\right.$$

$$+ \cos\left(\frac{2\pi N_1(S\omega/v)}{S\omega/v_0} + \frac{S\omega}{v} + \theta\right) + \cdots$$

$$\left.+ \cos\left[2\pi\left(\frac{N_1 + \cdots + N_n}{S\omega/v_0}\right)\frac{S\omega}{v} + \frac{S\omega}{v} + \theta\right]\right\} \qquad \text{(VII-16)}$$

where V_{RF} is the peak RF voltage, S is the grid spacing, θ is the phase angle of the RF as the ion passes the first grid of the analyzer, $v_0 = S\omega/2.3311$ is the velocity required for the ion to gain maximum energy, and N_n is the integral number of RF cycles necessary for a resonant ion to go between center grids of the nth and $(n + 1)$th analyzers.

Equation (VII-16) has a series of maxima depending on the number of stages and the number of RF cycles that separate each stage. However, within the arbitrary limits, $0.5 < (S\omega/v)/(S\omega/v_0)$ < 1.5 and $(S\omega/v + \theta) \approx \pi$, equation (VII-16) is nearly symmetrical about the value $(S\omega/v)/(S\omega/v_0) \approx 1$, which is the principal maximum. Hence the maxima will occur at

$$S\omega/v \simeq S\omega/v_0(1 \pm h_a) \qquad \text{(VII-17)}$$

where $(1 \pm h_a)$ is defined as the *harmonic ratio*, and the values of h_a depend on the values of $N_1, \ldots, N_n$. To a first approximation, values of $S\omega/v$ may be obtained from the maxima of a graphical plot of equation (VII-17) when

$$S\omega/v + \theta = \pi \qquad \text{(VII-18)}$$

The spectral peak which occurs for the maximum, $h_a = 0$ ($a = 0$), is called the fundamental peak of the ion mass M: The other spectral peaks occurring when $h_a = 0$ are called harmonics of the ion. Peaks associated with negative values of h_a are called the *upper harmonics*, and those with the positive values are the *lower harmonics*.

For a spectrometer with a voltage sweep, the values of V corresponding to these spectral peaks are determined by substituting v from equation (VII-17) into the energy equation

or
$$eV = \tfrac{1}{2} Mv^2 \qquad \text{(VII-19)}$$

$$V = M\left(\frac{S^2 f^2}{0.266 \times 10^{12}}\right)\left(\frac{1}{1 \pm h_a}\right)^2$$

where V is the potential in volts through which the ion has fallen in reaching the first grid of the analyzer stage.

A general case of a uniform flow of ionospheric ions of unknown mass and energy into an ion spectrometer with a voltage sweep can then be considered. Upon reaching the first grid of the analyzer, these ions will have fallen through $V = V_S + K$ volts, where V_S is the voltage applied to the analyzer by the sweep circuit and K is the unknown energy in volts which the ion possesses when entering the spectrometer system. Therefore, it follows that the fundamental and harmonic peaks of the unknown mass will occur at different but related sweep voltages. By subtracting any one of these voltages from another, the absolute mass of the unknown ion can be determined in terms of known parameters, irrespective of the unknown potential K. Thus, the mass of the unknown ion in terms of the voltage difference between its fundamental and a lower harmonic peak is given (in amu) by

$$M = \frac{0.266 \times 10^{12}}{S^2 f^2} \frac{V_S - V_{Sh_a}}{1 - (1/1 \pm h_a)^2} \qquad \text{(VII-20)}$$

Similar equations can be written for the fundamental and upper harmonic peaks, and lower and upper harmonic peaks. The quantity $0.266 \times 10^{12}/S^2 f^2$ is defined as the proportionality constant of the spectrometer.

The theory, as outlined here, presents a solution to the problem of determining the mass of an unknown ion. Values of the proportionality constant and harmonic ratio to be used in equation (VII-20) are readily obtainable experimentally for each spectrometer under actual operating conditions. The advantage of experimental determination of these parameters is that both constructional factors and circuit potentials, which were neglected to simplify the theory as presented, are automatically taken into account.

Since this method of mass identification makes use of harmonics in the mass spectra, the stopping potential applied to the spectrometer has to be reduced to permit the harmonics to appear. Experimentally this could be done by a stepping relay which changes the stopping potential in a cycle of six equal steps synchronized with the sweep, one step per sweep. The stopping-potential range is nominally adjusted to give three spectra with harmonics and three without. Application of this mass identification technique did solve several serious questions regarding upper-atmosphere ion-composition data obtained by rocket-borne spectrometers. The negative ion peak of mass 29 amu, which was reported as being a constituent of the daytime ionosphere at White Sands, New Mexico, was later refuted and identified as the lower harmonic h_3 of the very prominent mass 46 amu ion which had also been detected during that flight.

8. TWO-STAGE SINGLE-FREQUENCY RF ANALYZER FOR EXOSPHERE RESEARCH

At this stage, it seems appropriate to discuss the two-stage single-frequency RF mass spectrometer developed by C. Y. Johnson about 1960. This instrumentation has adequate sensitivity toward the 10 ions/cc density expected in the exosphere. It requires no critical adjustment of the DC and RF potentials and has low telemetry bandwidth requirements. Thus, it is most suitable to measure the hydrogen and helium content in regions around 1000 km. Also, in such a situation, high resolving power is not necessary. As altitude increases in the exosphere, the plasma becomes relatively cold and stationary, and the spectrometer encounters only ions of low energies.

The two-stage single-cycle spectrometer seems to be more suitable for ion measurements up to 1000 km with the required mass resolution, high ion-transmission efficiency, low "harmonic" level, and ion entrance-energy accommodation. The three-stage and four-stage single-cycle systems, though suitable for the job, have lesser efficiency in ion transmission and reduced ion entrance-energy accommodation.

Figure VII-14 shows in schematic form the tungsten mesh grid arrangement of a two-stage single-cycle spectrometer used for a laboratory prototype unit. Input grids 1 and 2 at negative potential draw ions into the spectrometer and electrostatically shield the analyzer from the ambient plasma. Grids 3 to 8 constitute the two-stage single-cycle mass analyzer in which the grids of each stage are equally spaced a distance S apart. The center grid of each stage is connected to an RF source through condenser C. A spacing of 0.7 S between grids 5 and 6 provides the proper one-cycle spacing required between identical grids of each stage. In this arrangement, an ion velocity v, which traverses the first stage in three-fourths of an RF cycle and crosses the RF grid as the field reverses to pick up maximum energy from the field, will then enter the second stage after traversing the drift space to cross the RF grid of that stage exactly one cycle later as the field reverses to gain again a second maximum increment of energy from the RF field. Ions of other velocities and phases are suppressed and decelerated. Resonant ions are maintained at a constant velocity within the analyzer.

Twin grids 9-10 and 11-12 are the retarding potential and suppressor grids, respectively, and they pass ions of only a particular energy level and beyond, suppressing the others. Secondary electrons emitted from the collector are returned to the retarding-potential grid by the suppressor.

The mode of operation of this instrument is entirely novel. The conventional single RF oscillator and linear accelerating potential mode of operation are discarded and instead a fixed accelerating potential in conjunction with four fixed-frequency oscillators which specifically analyze ions of mass numbers 1, 2, 4, and 16 are used. There is an automatic mass selector and, as a result, the mass peaks are not scanned but located in this system. The retarding potential is swept to obtain for each ion the information on its entrance energy into the spectrometer.

The maximum of 2.33 radians is the transit angle (α) required of the resonant ion in order that it gains the maximum energy in the Bennett system. α is the number of radians an ion of velocity v takes to traverse the distance S between adjacent grids of an RF stage for an applied radio frequency of $\omega = 2\pi f$ rad/sec. θ is the phase angle of the RF when the ion traverses the center grid of the first stage. For maximum energy gain, $\theta = \pi$ rad.

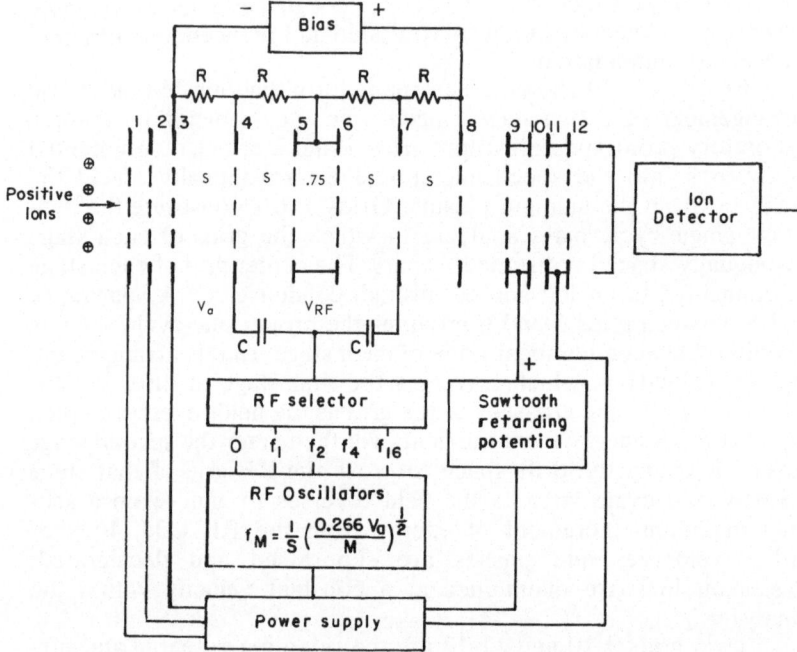

After C. Y. Johnson, from p. 1146. *Space Research—Vol. III* edited by W. Priester, North-Holland Publishing Co., Amsterdam (1963).

Fig. VII-14. Two-stage single-cycle radio-frequency mass spectrometer schematic used in Johnson's exosphere research.

The energy gain in volts for a two-stage single-cycle spectrometer is

$$V(\alpha, \theta) = -2\sqrt{2}\, V_{rf}\left(\frac{1 - \cos \alpha}{\alpha}\right)\left[\cos \theta + \cos(2.7\alpha + \theta)\right] \tag{VII-21}$$

where V_{rf} is the peak RF potential.

For all two-stage units, the resonant peak falls immediately to a minimum of zero energy-gain and there are harmonic peaks also. The maximum harmonic energy gain is 36% of the gain of the resonant ion. Operation of the spectrometer at the 40% retarding potential level when the harmonic ion currents are almost completely suppressed gives a theoretical ion transmission efficiency of about 15%, taking into account all grid-design parameters. Measured ion-transmission efficiency was about 50% of the expected value, which is an achievement.

Such a unit borne by a satellite and having an ion detector sensitive to 10^{-12} A will detect ion constituents with densities as low as 10 cm^{-3} when the spectrometer axis coincides with the velocity vector.

As designed, it is apparent that any ion greater than twice or less than half the mass of the resonant ion mass will not receive sufficient energy to pass through the potential barrier. This resolution of two is a fundamental requirement for the measurement of ions of oxygen, helium, and hydrogen. The system may be considered to be insensitive to ion entrance energy because of the broad-topped peak. However, the ion entrance energy itself acts to reduce the retarding potential, but this is readily compensated for by the saw-tooth voltage applied to the retarding grid. Figure VII-15 represents the spectra of some of the gases taken in the laboratory.

Johnson has given typical design parameters for a two-stage single-cycle RF ion mass spectrometer for investigating the earth's plasma. These parameters are as follows: Grid spacing, $S = 1.5$ cm; RF oscillator frequency, $f_m = (1/S)(0.266V_a/M)^{1/2}$; V_a, accelerating potential $= -300$ V; M is mass in amu; $V_{rf} = 5-10$ V; maximum energy gain is 4.1 V/V RF; bias voltage is about 3 V/V RF; retarding potential is approximately equal to a DC level of 40% of the maximum energy gain; sawtooth voltage amplitude is approximately equal to 60% of the maximum energy gain plus 15 V to compensate for ion entrance energy; size, about 500 in.^{3}; weight, about 5 lb; power, about 2 W.

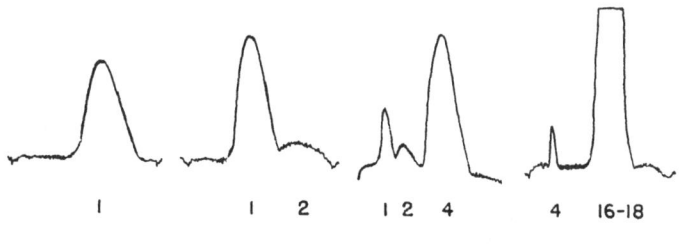

MASS AMU

After C. Y. Johnson, from p. 1149, *Space Research–Vol. III* edited by W. Priester, North-Holland Publishing Co., Amsterdam (1963).

Fig. VII-15. Laboratory spectra of helium and water vapor from a two-stage one-cycle RF mass spectrometer when operated with sawtooth accelerating potential and different radio frequencies.

9. RESULTS OF SOVIET EXPERIMENTS WITH RF MASS SPECTROMETERS

Before investigating the atmospheric composition studies conducted with the aid of quadrupole mass spectrometers and the very interesting composition data that has been obtained very recently, it is worthwhile to consider some of the composition research conducted by the Soviet scientists. By using RF mass spectrometers they obtained very interesting results concerning gravitational separation and mean molecular weight. In many cases their results contradict the results of other investigators; however, some of these discrepancies disappeared with subsequent experiments. The remaining controversies can be cleared up only by repeated experimentation and analysis.

a. Neutral Composition Above 100 km

Pokhunkov reported his investigations concerning the neutral composition at the First International Space Science Symposium in 1960. The investigations were conducted using a Bennett-type five-stage RF mass spectrometer which could analyze neutral gases in the mass ranges 1 to 4 amu and 12 to 56 amu. Ion currents from 6×10^{-13} to 8×10^{-9} A could be measured. The power requirement was 6 W and the weight of the apparatus was 3.3 kg. The altitude range from 100 to 200 km was investigated.

Gases with mass numbers 1, 2, 14, 16, 17, 18, 28, 32, 40, and 44 were registered and these numbers were identified as H_1, H_2, N_1, O_1, OH, H_2O, N_2, O_2, Ar, and CO_2, respectively. These results may be divided into two groups depending upon the character of the registered spectra. The first group is composed of H_2O, OH, H_2, and H_1 and the second of N_1, O_1, N_2, O_2, Ar, and CO_2.

The first group is characterized by a slow decrease of the ion-current amplitude with time during the flight. The amplitudes, however, do not depend upon the position of the analyzer axis relative to the velocity vector. The discovered excess of H_1 ($\approx 30\%$), beyond the amount which is in equilibrium with H_2O, is assumed to be the upper limit of the atomic hydrogen concentration in the altitude range 100 to 200 km and is equal to 1×10^8 atoms/cm^3.

The analysis of the ion-current behavior of the second group shows a characteristic regularity caused by container rotation during free flight. Therefore, one must consider the maximum values of the ion currents while evaluating the ion-current ratios since only then is the pressure change in the analyzer at a minimum. Some of the speculations of the investigator are reported here without an attempt to criticize them. The validity of the assumptions and speculations is determined when other experiments are also reported. Whenever it is possible and necessary, the apparent contradiction and relative validity of the results are pointed out.

When the ratio of atomic to molecular nitrogen was considered and referred to laboratory values, it was suggested that nitrogen, in the range 98 to 203 km, is composed of not more than 1 to 2% molecular nitrogen. The relative concentrations, O_1/N_2 and O_1/O_2, increase with altitude as shown in Fig. VII-16. However, the latter ratio grows more rapidly than the former and the relative concentration of O_2 in O_2/N_2 tends to decrease with altitude. Other results, such as the concentration ratio of carbon dioxide to molecular nitrogen, had to be discarded because of experimental errors and rocket influences.

The analysis of ion currents corresponding to Ar/N_2 shows gravitational separation of these gases above 100 km (see Fig. VII-17). However, not much could be said about this aspect by the investigators. It was also observed that the container was, on the

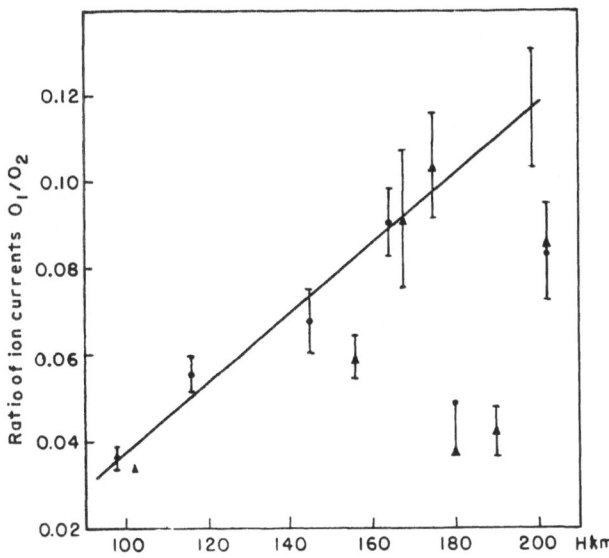

After A. A. Pokhunkov, from p. 105, *Space Research—Vol. I.* edited by H. Kallmann-Bijl, North-Holland Publishing Co., Amsterdam, (1960).

Fig. VII-16. The ratio of ion currents O_1/O_2 as a function of altitude. ▲ are the points taken on the outbound lap of the trajectory, ● those on the inbound lap.

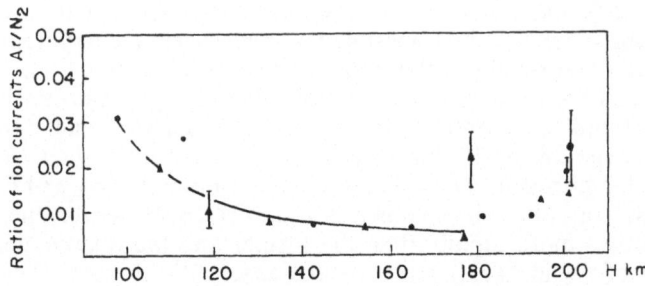

After A. A. Pokhunkov. from p. 106. *Space Research—Vol. I.* edited by H. Kallmann-Bijl. North-Holland Publishing Co.. Amsterdam. (1960).

Fig. VII-17. The ratio of ion currents Ar/N_2 as a function of altitude. ▲ *are points taken on the outbound lap of the trajectory,* ● *those on the inbound lap.*

average, charged negative up to a potential of almost 3 V above 90 km. This potential seems to indicate the effectiveness of the experiment for collecting positive ions.

b. Ion Composition from 230 to 1000 km by Means of the Third Soviet Satellite

A Bennett-type RF mass spectrometer was used in the third Soviet satellite to investigate the atmospheric ion composition. Daytime measurements were made in the range 225 to 980 km at latitudes from 27° to 65° N. The measurements showed that positive ions with mass number 16, probably atomic oxygen, predominate. Other positive ions with mass numbers 14, 18, 28, 30, and 32 were also recorded and were considered to be N^+, O^{18}, N_2^+, NO^+, and O_2^+. The peak intensity of atomic nitrogen relative to atomic oxygen seemed to change as a function of altitude and geographic latitude. Figure VII-18 shows the latitudinal variation of the intensity ratio. The observations made by the investigators are given in Table VII-3.

The ion-current ratios of NO^+ to O^+, O_2^+ to O^+, and N_2^+ to O^+ versus altitude are shown in Figs. VII-19, VII-20, and VII-21, respectively. The ion peak of molecular oxygen was observed up to a maximum height of 400 km and those of nitric oxide and molecular nitrogen up to a height of 500 km. The ion concentrations of NO^+ and N_2^+ become smaller and smaller with altitude and above 500 km it was speculated that the ionosphere consists only of atomic oxygen and atomic nitrogen.

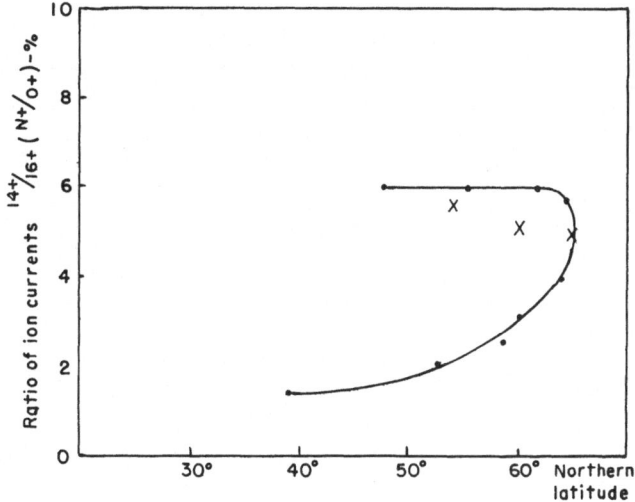

After A. A. Pokhunkov, from p. 109. *Space Research — Vol. I.* edited by H. Kallmann-Bijl. North-Holland Publishing Co., Amsterdam. (1960).

Fig. VII-18. Variation of i(14⁺/16⁺) with latitude, measurements taken by the third Soviet satellite. × Rev. 81, • Rev. 82, May 21, 1958.

c. Theoretical Developments

Consider, for a moment, the theoretical approach toward the problem of atmospheric composition in the Soviet Union about 1961. At this time exciting progress was being made in both the United States and the Soviet Union with the help of rocket-borne mass spectrometers. Various constituents of the atmosphere were identified and almost simultaneously reported. Istomin, Mirtov, Danilov, and others have contributed considerably toward the proper understanding of atmospheric composition with the available data.

In the Soviet Union the rockets with the mass spectrometers were usually claimed to carry the electronic devices necessary for the simultaneous measurement of electron density, n_e. Such an added feature is particularly significant because the combination of mass spectrometry data and the ion composition and electron density profiles enables the concentrations of ions of various masses to be determined very precisely as a function of altitude.

**Table VII-3. Spectrometer Information from
the Third Soviet Satellite**

Altitude range	Observation
225 to 350 km	Relative ion concentration of atomic nitrogen is considerably greater in latitudes 30 to 50° N than in 55 to 65° N
351 to 450 km	No latitudinal variation in the atomic nitrogen concentration
451 to 980 km	Latitudinal variation of atomic nitrogen becomes less important

The absolute ion concentrations were computed with the aid of the relation

$$[M_1^+] = \frac{n_e i_{M_1^+}}{\Sigma i_{M_k^+}} \qquad \text{(VII-22)}$$

which is valid if the following assumptions are fulfilled: (1) The atmosphere is electrically neutral, or

$$\Sigma[M_k^+] = n_e \qquad \text{(VII-23)}$$

which is based on the assumption that there are no negative ions in the atmosphere. Istomin claimed that this assumption may apparently be valid for the daylight ionosphere. (2) The sum of the amplitudes of the ion peaks on the mass spectrogram is proportional to the positive-ion concentration or

$$\Sigma i_{M_k^+} = k\Sigma[M_k^+] \qquad \text{(VII-24)}$$

This is equivalent to the absence of ions whose masses are not within the range of the instrument and, in this case, those masses in the range of the instrument are in the interval 1 to 4 amu and 10 to 56 amu. No serious doubts arise from this assumption. (3) The ratio of the amplitudes of the ion peaks in the spectrum is equal to the ratio of the corresponding ions in the atmosphere or

$$\frac{i_{M_1^+}}{i_{M_2^+}} = \frac{[M_1^+]}{[M_2^+]} \qquad \text{(VII-25)}$$

In other words there is no discrimination of ions of different masses in the mass spectrometer. Actually, this is not entirely true since there is apparently a certain amount of discrimination in the RF analyzer.

Istomin (1960) detected ions of several metals and estimated their concentrations with the aid of very sensitive RF mass spec-

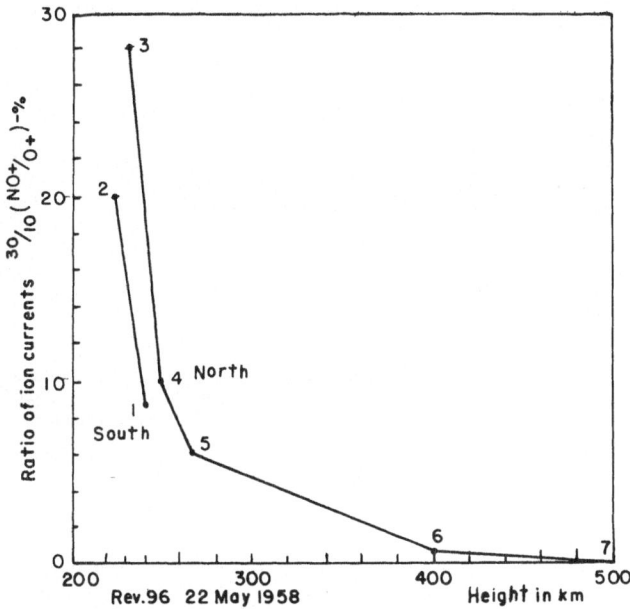

After A. A. Pokhunkov, from p. 110, *Space Research—Vol. I*, edited by H. Kallmann-Bijl, North-Holland Publishing Co., Amsterdam, (1960).

Fig. VII-19. i(NO⁺/O⁺) vs. altitude.

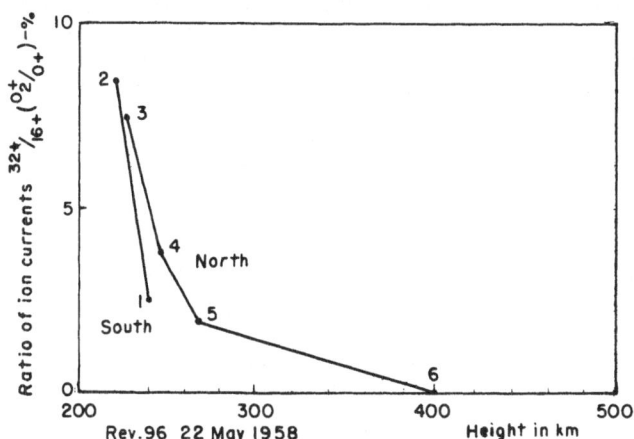

After A. A. Pokhunkov, from p. 111, *Space Research—Vol. I*, edited by H. Kallmann-Bijl, North-Holland Publishing Co., Amsterdam, (1960).

Fig. VII-20. i(O₂⁺/O⁺) vs. altitude.

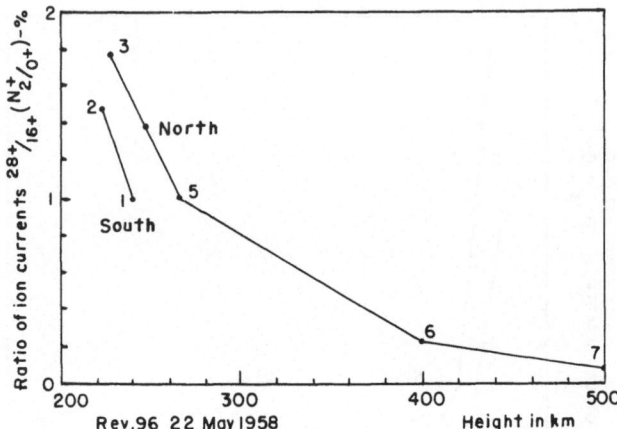

After A. A. Pokhunkov, from p. 112, *Space Research—Vol. I*, edited by H. Kallmann-Bijl, North-Holland
Publishing Co., Amsterdam, (1960).

Fig. VII-21. $i(N_2^+/O^+)$ vs. altitude.

trometers. In fact, calcium, magnesium, and iron ions were detected
and they seemed to form a narrow layer situated at 100 km with
a concentration of about $10^4/cm^3$. The thickness of the layer extended only a few kilometers. It has been reasonably assumed that
these ions are not of atmospheric origin and are a result of the decay
of meteoric matter that has penetrated the upper atmosphere.

These metallic ions are not dominant as far as composition
is concerned, but their presence is rather intriguing and challenging. In this context, Istomin suggested some production and loss
mechanisms which appeared to explain adequately the measured
compositions and molecular ion profiles.

Production: Ion exchange between atomic and molecular ions
is given as

$$A_1^+ + M_1 \rightarrow M_2^+ + A_2 \qquad \text{(VII-26)}$$

Loss: Dissociative recombination is given as

$$M^+ + e \rightarrow A_1 + A_2 \qquad \text{(VII-27)}$$

If the above mechanisms are applied to molecular nitrogen
then the formation, V_1, is given by

$$N^+ + N_2 \rightarrow N_2^+ + N$$

and the loss by

$$N_2^+ + e \rightarrow N + N$$

One may also consider the direct photoionization, V_3, of N_2 by the equation

$$N_2 + h\nu \rightarrow N_2^+ + e$$

Therefore,

$$V_1 = (N^+) \cdot (N_2) \cdot \gamma \qquad (VII\text{-}28)$$

and

$$V_3 = (N_2)\sigma n \qquad (VII\text{-}29)$$

where γ is the rate coefficient for charge transfer reaction, 10^{-10} cm^3/sec; σ is the photoionization cross section, 10^{-17} cm^3, and n is the number of quanta of radiation per cm^2-sec.

If n is taken to be the value suggested by Byram, Chubb, and Friedman, then $V_1/V_3 \approx 10^2$. If $n = 6 \times 10^{11}$ quanta/cm^2-sec, obtained by Ivanov-Kholodny for altitudes greater than 250 km, then $V_1/V_3 \approx 10$.

d. Nine Rocket Soundings in the European Sector of the USSR

At the Third International Space Science Symposium, Istomin reported the results of nine rocket soundings on atmospheric composition. All nine rockets carried Bennett-type RF mass spectrometers. Table VII-4 presents information about the rockets and the mass spectrometer instrumentation.

These measurements were made in the altitude range 110 to 210 km and were performed by the method of separated containers which was being widely used in the USSR. The rocket, a principal source of atmospheric pollution, did not affect the measurements during a considerable part of the trajectory because during the ascent the container with the neutral and ionic mass spectrometers was ahead of the rocket at a distance equal to many mean free paths of a molecule. Ion rocket pollution was recorded by the ionic mass spectrometers mainly on the descending part of the container trajectory. The instrument was extremely sensitive.

Istomin also reported on the third Soviet satellite (*Sputnik III*) where, for the first time, atmospheric composition at heights of 225 to 980 km was studied by an RF mass spectrometer. However, it was argued by many scientists (Newell, etc.) that the continuous strong release of gas from the satellite surface might have polluted the neighboring atmospheric regions and hence the mass spectrometric observations might not have given correct information. Mirtov argued to the contrary and the *Sputnik* measurements seem to support his views. Actually, pollution ions are easily distinguishable in the mass spectrum. The H_2O^+ pollution ions are probably

Table VII-4. Instrumentation Used in Russian Rocket Soundings Reported at the Third International Space Science Symposium

Name of the instrument	PMC-1	MX-6401	MX-6403
Number of cycles in the analyzer drifts	5-7	5-9-4-7	5-9-4-7
Power demand (W)	30	5.3	3.2
Weight (kg)	6.5	3.3	2.0
Mass scale (amu)	6 to 48	1 to 4	1 to 4
	8 to 63	12 to 56	12 to 56
Resolution – maximum	30	50	50
– normal	20	20	20

formed as a result of interchange processes of the pollution molecules with atomic oxygen ions.

$$H_2O + O^+ \rightarrow H_2O^+ + O \qquad \text{(VII-30)}$$

i. Ion Composition. The main results for the mass spectrum of the positive ions at heights of 100 to 210 km were obtained as mentioned earlier, in five rocket soundings, at moderate latitudes in the European sector of the USSR. The results of the flight on September 9, 1957, revealed the dominating characteristic of the ions of mass number 30, which were attributed to nitric oxide, this being confirmed by other flights.

Various other ions were recorded and identified by the investigators as follows: 32^+ as O_2^+; 16^+ as O^+; 14^+ as N. The last two are minor atmospheric constituents at 100 to 200 km. That these minor constituents could be clearly recorded was contrary to the results of C. Y. Johnson at the First Space Science Symposium.

The main components NO^+, O_2^+, and O^+ showed conspicuous diurnal variations. The relative atomic oxygen concentration at all heights from 140 to 200 km showed a rapid increase with the rising sun. However, the relative concentration of O_2^+ decreased during the daytime. The height of maximum concentration of O_2^+ varied from 160 km in the morning to lower heights in the evening. Calculations of the absolute ion densities were made according to equation (VII-22).

Figure VII-22 shows the positive ions of atomic and molecular nitrogen as a function of altitude. N_2^+ ions are concentrated in the ionospheric regions in a range between 250 and 350 km with maximum concentration equal to 1×10^4 cm^3 at 250 km. On the other hand, atomic nitrogen ions appear in appreciable quantities (more

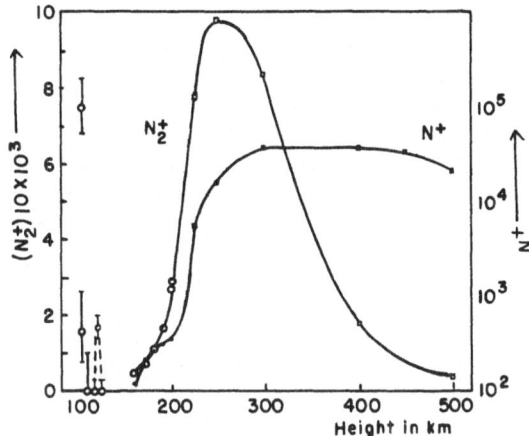

After V. G. Istomin and A. A. Pokhunkov, from p. 125, *Space Research—Vol. III*, edited by W. Priester, North-Holland Publishing Co., Amsterdam, (1963).

Fig. VII-22. Distribution of the positive atomic and molecular nitrogen ions in the earth's atmosphere according to rocket data obtained on August 2, 1957, June 15, 1960, and the third Soviet satellite, May, 1958.

than 100 ions per cm³) above 160 km. Initially, the absolute concentration of N⁺ seems to increase rapidly, reaching a maximum at 300 km, and thereafter remains constant or decreases slightly. This variation is unique.

The scale height, $H = kT/Mg$, for $M = 14^+$, appeared to be rather large in the height range 300 to 500 km and this seemed to indicate that the nitrogen ions ($M \approx 14^+$?) are not in thermodynamic equilibrium with the neutral atmospheric components. The N⁺ ions therefore have higher thermal velocities than the neutral components. This reasoning was qualitatively confirmed by the dependence of the N⁺ and O⁺ ion current components on the angle between the mass spectrometer tube and the satellite velocity vector. The corpuscular hypothesis of ionization by electrons of comparatively low energy also predicts higher thermal velocities for N⁺ and O⁺.

For nitrogen, $N + e \rightarrow N^+ + 2e$, the relative energy of the atoms is zero only when the electron energy is 24.5 eV. However, since electron energy extends up to hundreds of electron volts, the atomic nitrogen ions should be considerably energetic. The "hot" ions can diffuse up to higher altitudes than "ordinary" ions and thus increase the scale heights. "Hot" ions must also exist in the earth's exosphere.

ii. Neutral Gas Composition and Gravitational Separation.
The most important result which defines the upper atmospheric
conditions is the detection of gravitational separation between argon
and molecular nitrogen. Very interesting results were obtained by
Istomin and Pokhunkov in their rocket experiments which were
conducted in the middle latitudes in the European sector during
1960 and 1961. The separation level was located at a height of 105
to 110 km. Townsend *et al.* obtained similar results at polar latitudes (Fort Churchill). The two rocket soundings used to observe
gravitational separation and neutral gas composition were conducted in September 1960 (night) and November 1961 (day). Actually, the neutral gas composition measurements were carried out
after the separation of the container at distances up to a few hundred meters from the rocket.

A very peculiar phenomenon observed during these experiments was the periodic modulation of the ion currents produced
by an arbitrary container rotation in its flight and by a pressure
head effect. The spectra recorded ion peaks corresponding to gases
of the following mass numbers: 1, 2, 12, 14, 16, 17, 18, 28, 29, 30,
32, 34, 36, 40, 42, and 44. These numbers were identified with H_1,
H_2, C, N, O, OH, H_2O, N_2, $N^{14}N^{15}$, NO, O_2, $O^{16}O^{18}$, $Ar^{36}Ar^{40}$,
$Mg^{26}O$, CO_2, and N_2O, respectively.

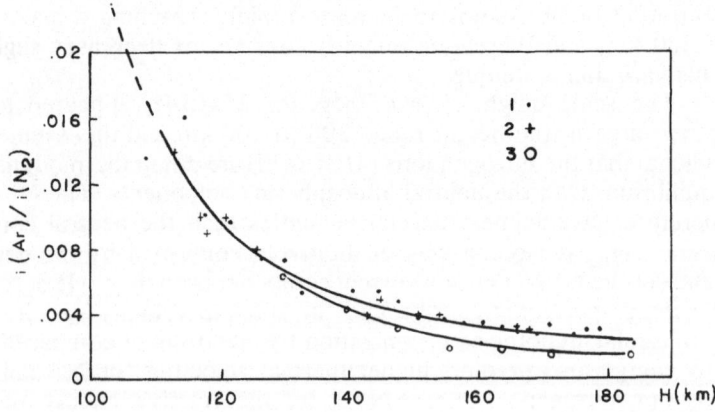

After A. A. Pokhunkov, from p. 134, *Space Research—Vol. III*, edited by W. Priester, North-Holland Publishing Co., Amsterdam, (1963).

*Fig. VII-23. Variation of the Ar and N_2 ion current ratio with height. 1 and
2—ascent and descent of the flight (night, 1960); 3—ascent (day, 1961).*

Diffusive separation of Ar and N_2 was noticed in both the day and nighttime atmosphere and seemed to be explicable in the variation of the Ar and N_2 ion current ratio with height. The altitude range was fixed around 110 to 120 km. Figure VII-23 shows the current ratio versus altitude. Since the gravitational separation exists in the altitude range of interest, the equation for scale height used to calculate atmospheric temperature was slightly altered in calculating the temperatures up to 185 km. The altered equation for the scale height is

$$H = \frac{RT}{(M_{Ar} - M_N)g} \qquad \text{(VII-31)}$$

During the daytime experiment of 1961, when the instrument orientation was constant, the temperature was computed from the slope of the curve for N_2 ion current with a scale height given by

$$H = H_o \frac{2h_m - h}{(2h_m - h) - H_o} \qquad \text{(VII-32)}$$

where H_o is the scale height for the N_2 concentration inside the mass spectrometer analyzer obtained from the steepness of the slope of the N_2 ion current curve, $h_m = 430$ km, the maximum height of ascent, and h is the height at which H_o is measured. The measured values of temperature and density of the various constituents are presented in Figs. VII-24 and VII-25.

There were many interesting results obtained regarding the neutral gas composition by Istomin and Pokhunkov. One interesting aspect of the results was that neutral particles with the mass number 42 were detected from 100 to 120 km and were thought to be magnesium oxide. It was also speculated that ions, such as Mg^+, Ca^+, Fe^+, and Si^+, of meteoric origin, might be of some importance in the earth's atmosphere. However, scarcity of substantial experimental evidence prevented the investigators from feeling assured of the existence and origin of these ions and their best effort was a speculation.

It was shown that a nitrogen–oxygen atmosphere exists at least up to a height of 300 km. It was also shown that molecular nitrogen remains a dominant atmospheric constituent up to heights of 280 km, where the concentrations of molecular nitrogen and atomic oxygen are approximately equal. It is also important to note that the dissociation of molecular nitrogen was unimportant up to an altitude of 210 km. At heights exceeding 100 km, atomic oxyger was experimentally detected and $n(O)/n(N_2)$ increases gradually

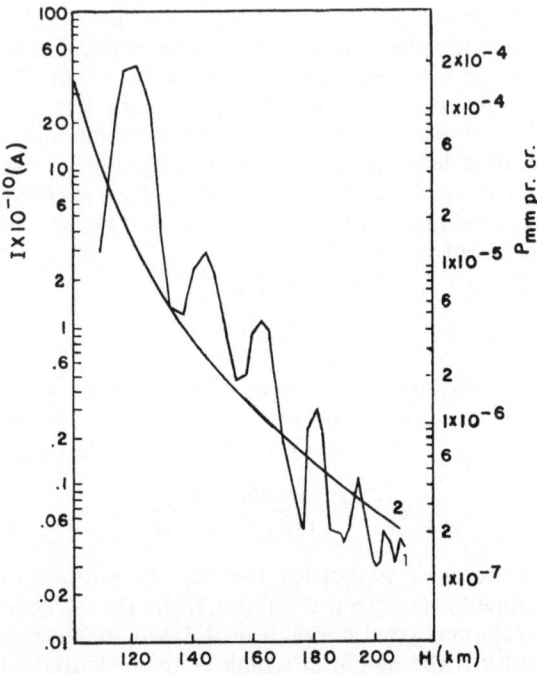

After A. A. Pokhunkov, from p. 138, *Space Research—Vol. III*, edited by W. Priester, North-Holland Publishing Co., Amsterdam, (1963).

Fig. VII-24. 1—Variation of the N_2 ion with height during the ascent with different orientation of the mass spectrometer (night, 1960). 2—N_2 pressure variation in the atmosphere according to the barometric formulae.

with altitude thereafter. One of the experiments gave the following results:

$$\frac{n(O)}{n(N_2)} \approx 0.65 \pm 0.20 \qquad\qquad (VII\text{-}33)$$

and

$$\frac{n(O_2)}{n(N_2)} \approx 0.14 \pm 0.06 \qquad\qquad (VII\text{-}34)$$

at 210 km. Nevertheless there is a considerable amount of undissociated oxygen at 210 km, which indicates the essential importance of diffusion in restoring the molecular oxygen height distribution disturbed by photochemical reactions.

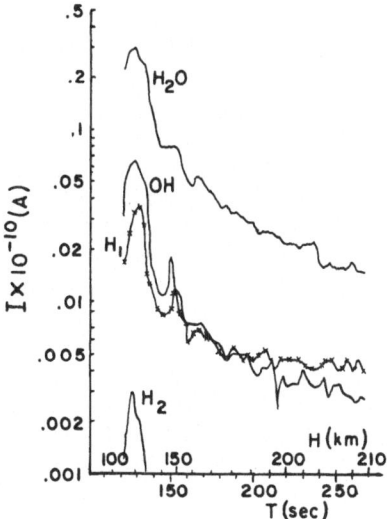

After A. A. Pokhunkov, from p. 140, *Space Research – Vol. III*, edited by W. Priester, North-Holland Publishing Co., Amsterdam, (1963).

Fig. VII-25. Variation of the H_2O, OH, H_1 and H_2 ion currents with height (time) on the ascent of the flight.

The investigators claimed that temperature estimates by means of a mass spectrometer are more accurate than those made from manometric data since the latter requires an exact determination of the altitude variation of mean molecular weight. A study of atmospheric composition by means of a mass spectrometer at altitudes where chemically active atomic components exist, which can partially combine or recombine with the material on the inner walls of the analyzer, presents great difficulties. On the basis of comparison between the experiments performed with mass spectrometer analyzers of different designs, a correction factor was determined which takes into account possible reactions inside the analyzer which change the analyzing-gas composition (see Appendix). The experimenters observed that within the limits of apparatus sensitivity, when the analyzers were opened, neutral helium was not recorded in the spectra. This confirms the fact that at altitudes above 100 km the helium concentration does not amount to more than 6×10^7 particles per cm³.

iii. H_2O, OH, H_1, and H_2. According to the nighttime experimental data, the H_2O ion current decreases with altitude. The maximum value of the H_2O partial pressure in the upper atmosphere, estimated according to the ion-current modulation depth, does not exceed 3×10^{-7} mm Hg at 115 km or 0.6% of the total atmospheric pressure. This value may be the upper limit of the H_2O content in the atmosphere, since ion-current modulation can be produced by the number of reflected molecules of H_2O entering the analyzer when the analyzer changes its orientation relative to the contrary flow. If it is assumed that the OH content recorded by the mass spectrometer is produced by H_2O dissociation in the ion source of the apparatus, then the total OH content above 100 km is not greater than 6×10^{-3}% of the total atmospheric pressure. The experiments justify this argument. The H_1 ion-current pattern *versus* altitude is analogous to the H_2O and OH patterns and this justifies the dissociative relation of H_1 and H_2O. The recombination of atomic hydrogen inside the analyzer fully accounts for the formation of the molecular hydrogen found at altitudes up to 130 km, as is evidenced by the H_1 ion-current correlation. Therefore, in the absence of H_2 in the spectra above 130 km, the upper limit of H_2 concentration in the atmosphere above 100 km is set at 3×10^7 particles per cm^3.

iv. Detection of Metallic Ions. During the nighttime experiment, molecules with mass number 42 were detected at 103 to 126 km. Istomin suggested that this might be $Mg^{26}O^{16}$. The absolute concentration of all magnesium oxides obtained by considering their relative distribution was given as 10^9 particles per cm^3 in the altitude range 103 to 126 km. Stone meteors, in which the MgO content could be up to 40.2% of the total weight, seem to be the apparent source of MgO. Stone meteors comprise a major portion of all the meteors shooting into the atmosphere. MgO might be formed as a result of direct meteor evaporation or oxidation of Mg atoms in the atmosphere. The absence of metallic magnesium in the spectra and the presence of sufficient atomic oxygen in the altitude range of interest strongly upholds the effectiveness of the oxidation process. The decrease in the relative concentration of $Mg^{26}O$ above 117 km seems to agree with the observations made with respect to the gravitational separation of gases at altitudes above 105 to 110 km. The decrease of $Mg^{26}O$ below 117 km could be explained in terms of a transition layer, 105 to 110 km, where the gravitational separation, which is just beginning, is being partially disturbed by separate flows from the turbulent atmosphere lying below. Thus, due to turbulent mixing, the MgO content falls down.

10. COMPOSITION STUDIES WITH QUADRUPOLE AND TIME-OF-FLIGHT MASS SPECTROMETERS

Better and better techniques have been developed as the investigations of atmospheric composition have advanced. The investigations started with the postflight analysis of upper-air samples obtained in containers completely detached from the rocket while in flight. The RF mass spectrometers were introduced in 1955.

In 1961 Schaeffer and Nichols carried out an experiment with Paul's quadrupole mass spectrometer for in-flight composition analysis. The analyzing section of the mass filter consisted of four circular rods, positioned 90° apart and connected as shown in Fig. VII-26. The applied potentials are given by

$$V_{0x} = U + V \cos \omega t \qquad (VII-35)$$

and

$$V_{0y} = - V_{0x} \qquad (VII-36)$$

where V is the peak value of an RF sine wave superimposed upon the DC voltage U. Ideally, the rods have a hyperbolic cross section in which case the potential at any point in the field is given by

$$V_{x,y} = (U + V \cos \omega t)\frac{x^2 - y^2}{R_o^2} \qquad (VII-37)$$

In practice, circular rods of radius 1.16 R_o closely approximate the ideal field out to about 0.8 R_o.

The equations of motion of a singly charged ion of mass m in such a field are

$$m\ddot{x} + 2e(U + V \cos \omega t)\frac{x}{R_o^2} = 0 \qquad (VII-38)$$

and

$$m\ddot{y} - 2e(U + V \cos \omega t)\frac{y}{R_o^2} = 0 \qquad (VII-39)$$

Letting $p = \omega t/2$, $a = 8eU/mR_o^2\omega^2$ and $q = 4eV/mR_o^2\omega^2$, the equations of motion become

$$x'' + (a + 2q \cos 2p)x = 0 \qquad (VII-40)$$

and

$$y'' - (a + 2q \cos 2p)y = 0 \qquad (VII-41)$$

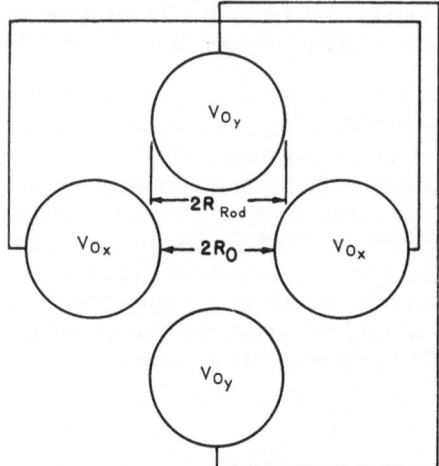

After E. J. Schaeffer and M. H. Nichols, *ARS J.* **31**, 1773, (1961).

Fig. VII-26. Circular electrode arrangements.

where the primes denote differentiation with respect to p. These are the Mathieu differential equations and the general solutions, including a general analysis of mass filters, are discussed in Chapter VI. The solutions of equations (VII-40) and (VII-41) in terms of a and q yield regions where the ion trajectories are either stable or unstable. A stable trajectory is effected if both x and y solutions are stable. The values of a and q at which this condition occurs are given in Fig. VII-27.

If the ratio of DC to RF voltage is held constant, a/q plots as a straight line in Fig. VII-27. This line is the locus of the working points of all ions.

Heavy ions plot close to the origin and lighter ions plot progressively toward the upper right. If the voltages are increased from zero, the working points will move away from the origin. The lightest ion will pass through the stable region first and will be followed by the heavier ions in the order of increasing masses. Ideally, if a stream of ions is injected at one end of the field, unstable ions are removed by collisions with the rods. Stable ions negotiate the length of the field and are detected when they deliver their charges to a collector placed at the other end. In practice, however, there is a maximum number of cycles of the applied RF voltage for which an unstable ion must remain in the field to ensure removal. Furthermore, the initial conditions of radial velocity and displacement must

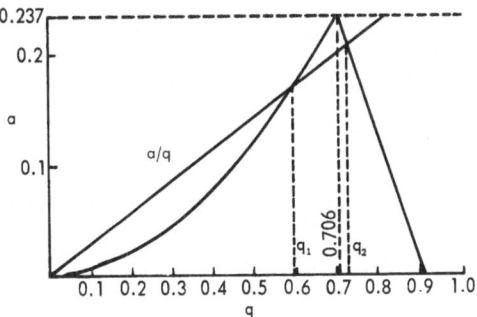

After W. Paul *et al., Z. Physik* **152,**146, (1958); also E. J. Schaeffer and M. H. Nichols,*ARS J.***31,** 1773, (1961).
Courtesy of Springer-Verlag, Berlin.

Fig. VII-27. Massenfilter stability diagram.

be restricted to ensure collection of a stable ion. However, both restrictions increase in severity with resolution.

At low RF voltages all ions are in the stable region. As the RF voltage is increased the working points move away from the origin so that ions become unstable in order of increasing mass.

The mass filter was selected by Schaeffer and Nichols because of the following advantages: (1) It requires no magnetic field, hence, it is inherently light in weight; (2) its construction is simple yet rugged; (3) initial conditions affect only percentage transmission and resolution and have no effect on the indicated mass of the ion; (4) supporting circuitry is simple. A constant voltage ratio is maintained by rectifying a portion of the RF voltage to obtain the DC voltage; (5) a linear mass spectrum is obtained from a linear voltage sweep.

For design purposes, resolution, peak RF voltage, rod length, and injection voltage were suitably selected by the experimenters. Having been defined as the mass of the ion divided by the width (expressed in mass units) of its spectral peak at half-amplitude, a resolution of 40 for mass 46 was considered adequate to separate the gases that could reasonably be expected in the upper atmosphere. Because of the shorter residence time of the lighter ions, resolution would become less as the mass is reduced but less resolution is required as mass decreases. The peak RF voltage was selected to be 500 V and was conveniently obtained in a miniaturized flight package. Rod length was 12.75 cm, a reasonable size for small rocket-borne experiments. Finally, an injection ion energy of 45 V was selected as a compromise between a low ion velocity that reduces RF power requirements and a high ion energy that avoids the difficulties inherent in controlling low-energy particles.

Table VII-5. Design Parameters for Mass Filter Selected by Schaeffer and Nichols

Quantity	Symbol	Value
Mass number	A	46 amu
Resolution	$M/\Delta M$	40
Peak RF voltage	V	500 V
Rod length	L	12.75 cm
Ion injection voltage	V_{in}	45 V
RF frequency	f	2.39 Mc
Field radius	R_o	0.522 cm
Rod radius	R_{rod}	0.609 cm
Injection port diameter	D_{in}	0.081 cm
Maximum injection angle	θ	5.25°

The important design parameters are summarized in Table VII-5. A laboratory model incorporating all these parameters is shown in Fig. VII-28. Resolution of the nitrogen peak as a function of the ion accelerating voltage is presented in Fig. VII-29.

Perhaps the largest uncertainty in the previous measurements was the error in analysis caused by recombination of atomic oxygen on the walls of the gauge. Another uncertainty was the contribution of occluded gases on the rocket. Contamination by rocket gases was avoided in this experiment by ejecting the entire pressurized instrument, as the Russians were doing with RF mass spectrometers, from an evacuated volume when the desired altitude range was reached. Emerging from the vacuum the surface of the instrument cylinder was clear of occluded gases.

Recombination was minimized by an ion source composed of fine grid structure directly immersed in the ambient gas. Such an arrangement increases the probability that an ambient molecule would enter the ionization volume without first encountering a solid surface where recombination could occur. However, the best way to avoid this defect would be to design an improved ion source which is highly efficient.

Narcisi *et al.* have been successfully carrying on composition measurements with the aid of time-of-flight and quadrupole mass spectrometers. The first development of their investigations was evident when they reported a mass spectrometer suitable for use in rockets for nitrogen and oxygen atom sensitivities at the Second International Space Science Symposium.

Active gas atoms of nitrogen and oxygen were produced in a flow system by subjecting molecular nitrogen to a microwave discharge. The method of producing known concentrations of N and

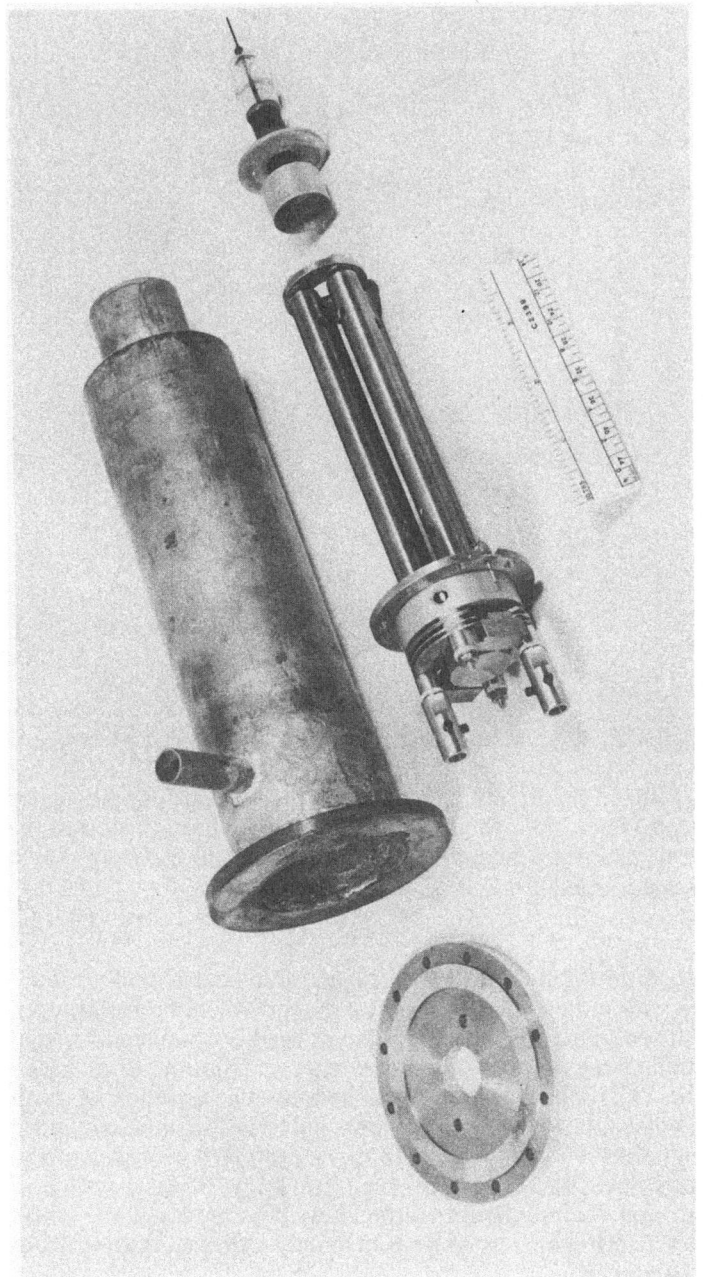

After E. J. Schaeffer and M. H. Nichols, *ARS J.* **31**, 1773, (1961).

Fig. VII-28. Laboratory massenfilter.

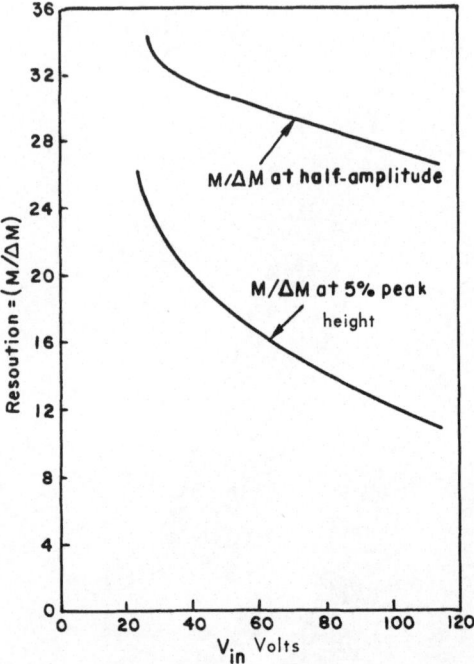

After E. J. Schaeffer and M. H. Nichols, *ARS J.* 31, 1773. (1961).

Fig. VII-29. Resolution of nitrogen as a function of injection voltage.

O atoms utilized the following reaction, which is considerably faster than any other plausible chemical reactions when the concentrations of N, O, and NO are of the order of 10^{14} molecules per cm^3. The reaction is given as

$$N + NO \rightarrow N_2 + O \qquad \text{(VII-42)}$$

In this reaction the flow rate of nitrogen atoms could be "titrated" with nitric oxide molecules to produce an equivalent flow of oxygen atoms. The end point of this reaction was readily obtained by visual observation of the light emission downstream from the nitric oxide inlet. The end point was distinguished by the absence of both greenish-yellow (excess nitric oxide) and blue (insufficient nitric oxide) emissions. Only then is the oxygen atom flow equal to the nitric oxide flow. By shutting off the nitric oxide it was possible to obtain an equivalent nitrogen atom flow. By reducing the nitric oxide flow a nitrogen–oxygen atom mixture of any desired ratio could be obtained.

The gas in the flow system was continuously sampled into the mass spectrometers by means of a leak. This leak was constructed from a Pyrex thimble about 20-μ thick in which a 20-μ hole was punched. At the prevailing flow pressures the hole diameter was just sufficient to ensure molecular flow. A Bendix time-of-flight mass spectrometer that was used by the experimenters is shown in Fig. VII-30 and VII-31.

After the neutrals are ionized in the spectrometer the masses are separated by giving all ions in a bunch the same energy and allowing all ions to traverse the same distance. The time of flight for an ion of mass m is

$$t = \frac{S}{(2eV/m)^{1/2}} \qquad \text{(VII-43)}$$

where V is the accelerating voltage and S is the flight distance. The lighter masses have a shorter time of flight.

Electrons enter the ionization chamber of the mass spectrometer when the control grid is pulsed at a 10 kc rate to the filament potential. An energy of 400 eV is imparted by the ion energy grid to all the ions which then travel down the drift tube. By putting the basic timing pulse into a delay line which has multiple taps, delay times can be selected corresponding to arrival times of the ions of different mass at the electron multiplier. Ion currents of a particular mass are then amplified and collected at an anode by suitable gating pulses generated by a signal from a delay tap. The spectrometer was

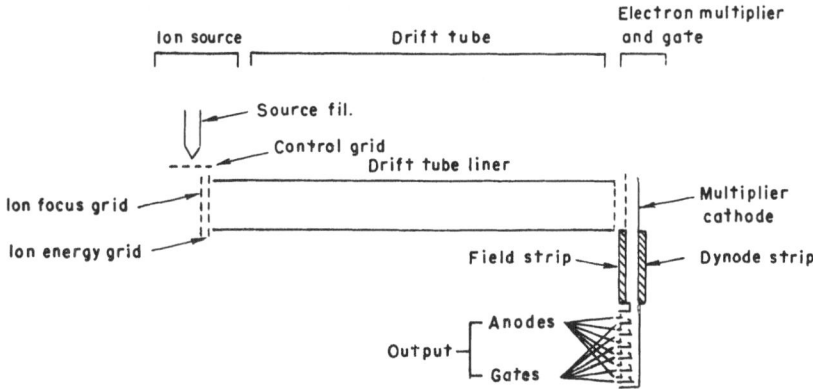

After R. S. Narcisi *et al.*, from p. 1157, *Space Research — Vol. III,* edited by W. Priester, North-Holland Publishing Co., Amsterdam, (1963).

Fig. VII-30. Mass spectrometer physical system (Narcisi).

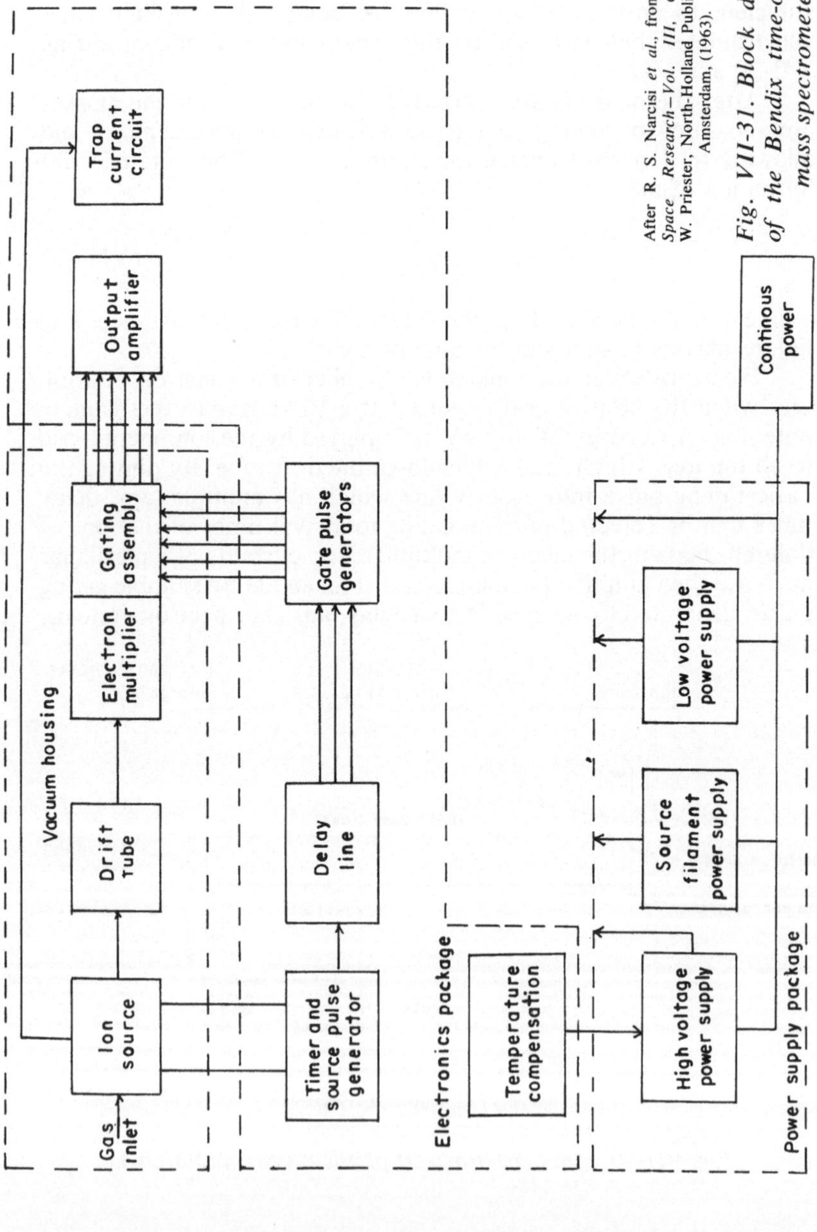

After R. S. Narcisi *et al.*, from p. 1158, *Space Research—Vol. III*, edited by W. Priester, North-Holland Publishing Co., Amsterdam, (1963).

Fig. VII-31. Block diagram of the Bendix time-of-flight mass spectrometer.

designed in such a way as to measure six mass peaks simultaneously. Spectrometer pressure was maintained at 2×10^{-5} torr and the spectrometer itself had a volume of 1 liter.

Figures VII-32 and VII-33 show two different types of spectrometers that were investigated. Figure VII-32 has a head-on entrance port designed for oriented satellites and Fig. VII-33 has a backing plate designed for tumbling satellites. The features of these two types are listed in Table VII-6 with regard to neutrals entering the ionization region.

The experimenters reported an attempt to measure the spectrometer efficiency for nitrogen and oxygen atoms with head-on geometry at an electron ionization energy of 91 eV. They noticed no change in the currents at masses 14 (N^+) and 16 (O^+) between discharge off and on, but there was a change in the current at mass 30 under similar conditions. This observation is significant considering the fact that the ionization efficiencies of N and O are comparable to that of NO at an electron energy of 91 eV.

When the experiment was repeated with the ion repeller grids out and using 18 V electrons, significant changes in current at mass 14 (N^+) and considerably smaller changes in current at mass 16 (O^+) were detected with and without discharge. When the experiment was again repeated with the backing plate geometry near 18 V ionization energy, the results corresponding to mass 14 (N^+) were similar to those obtained with head-on geometry. There appeared to be a significant change in the current at mass 16 (O^+).

It was observed that the optimum detectability for a change in the current at mass 14 (N^+) was at an ionization energy below 21 V. It was suggested that this could be caused by the high dissociative ionization background due to the flowing N_2.

Some problems encountered in the design and calibration of the instrument provide a broader insight into some of the intricate problems which are not easily seen. This is important from the point of view of the development of instrumentation. Some of the significant conclusions arrived at by Narcisi *et al.* were: (1) The spectrometer sensitivity for nitrogen atoms is about an order of two less than that for nitric oxide molecules at an ionization energy of 18 eV. (2) The spectrometer sensitivity for nitrogen atoms remains unaffected when the number of wall collisions increases. (3) The concentration of nitrogen atoms in the presence of a strong background of nitrogen molecules can be measured accurately only when the ionization voltage is suitably chosen to reduce the background produced by dissociative ionization. (4) The spectrometer sensitivity for oxygen atoms is at least a factor of five less than that for nitrogen atoms and depends upon the number of wall collisions.

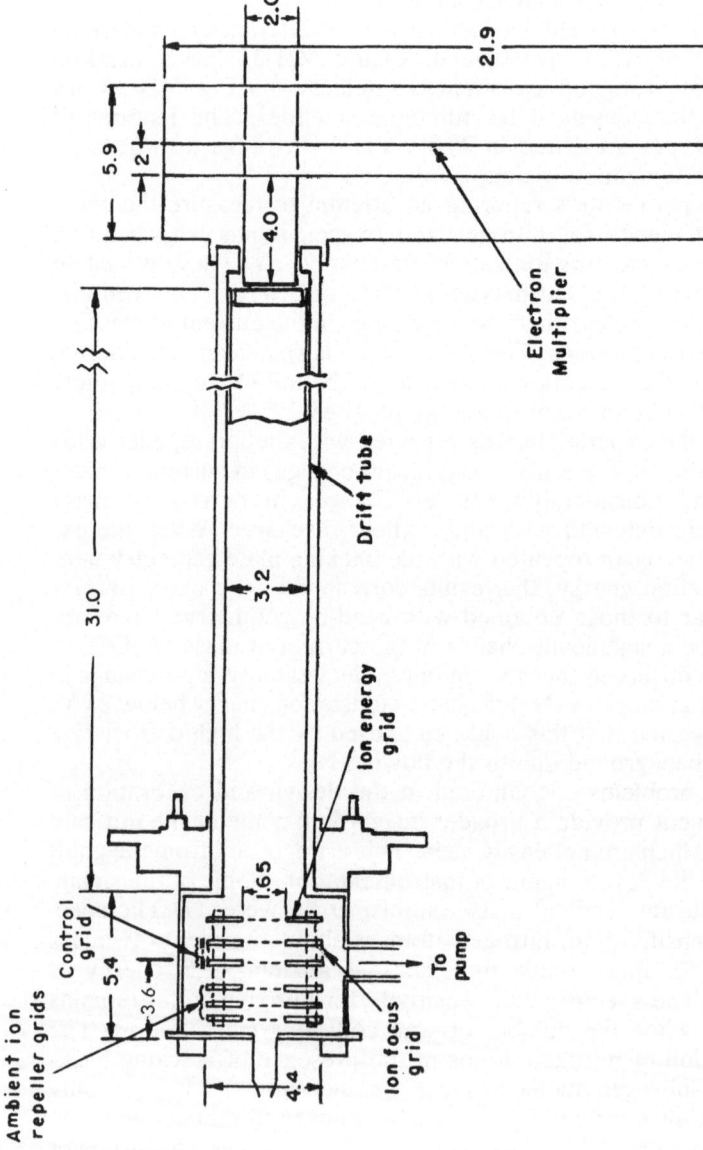

After R. S. Narcisi et al., from p. 1160, Space Research—Vol. III, edited by W. Priester, North-Holland Publishing Co., Amsterdam, (1963).

Fig. VII-32. Physical dimensions of the satellite time-of-flight mass spectrometer showing the leak system used during calibration. Head-on entrance chamber type. All dimensions in centimeters.

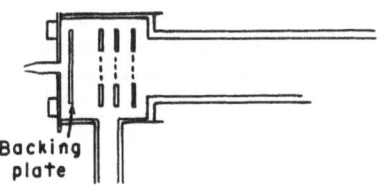

Backing
plate

After R. S. Narcisi *et al.*, from p. 1160, *Space Research — Vol. III*, edited by W. Priester, North-Holland Publishing Co., Amsterdam, (1963).

Fig. VII-33. Backing plate entrance chamber type.

11. A QUADRUPOLE MASS SPECTROMETER FOR *D*- AND LOWER *E*-REGION COMPOSITION MEASUREMENTS

Narcisi *et al.* successfully flew a quadrupole mass spectrometer system on a Nike–Cajun rocket on October 31, 1963, at Eglin Air Force Base, Florida, and thus made the first ion-composition measurements in the *D* region. The rocket attained a peak altitude of 111.7 km. Up to this time no comprehensive environmental measurements of the ion composition had been performed below 90 km. Thus, the experimenters have opened a new era in the understanding of the chemical dynamics of the *D* region. The mass spectrometer used employed a liquid-nitrogen chilled Zeolite pump which made it possible to measure ion composition of such low altitudes. On this same rocket flight, simultaneous measurements of ion and electron densities were made using two electrostatic probes. Therefore it was possible to study the ionic content of the ionosphere with a greater degree of confidence.

The mass spectrometer system was set up just beneath the nose tip (with vacuum cap) of the rocket. The nose tip and vacuum cap had been programmed to be separated and ejected from the rocket in the vicinity of 60 km. Figure VII-34 shows a cross section of the front end of the payload in the sampling configuration without nose tip and vacuum cap. The entire spectrometer package is electrically insulated from the rocket in order to facilitate proper biasing of the instrument package to collect positive ions. The design and operating parameters of the mass spectrometer constructed of stainless steel rods are given in Table VII-7.

A higher sensitivity was obtained by using a higher ion injection voltage but at the expense of mass resolution. However, a value of 16 for resolution was found satisfactory. Since the experiment was designed to investigate the ion composition of the *D* region, the investigators had to estimate the instruments's sensi-

Table VII-6. Features of Two Spectrometers with Regard to Neutrals Entering the Ionization Region

With Head-on Entrance Port	With Backing Plate
(1) Direct entrance without colliding with the walls	(1) No direct entrance
(2) Entrance by diffusion around the entrance structures	(2) Entrance by diffusion around the entrance structures
(3) Entrance by back-diffusion after entering the multiplier region	(3) Only nonionized molecules can diffuse back from the multiplier region to the ionization region

tivity around 60 km. The difficulty of simulating such a weak plasma in the laboratory made the experimenter's estimate highly doubtful. Therefore the instrument was suitably designed to receive small ion currents. The investigators did not have the advantage of utilizing the sensitivity calculations made from previous D-region rocket flights since theirs was the first. The electrical block diagram of the spectrometer system is shown in Fig. VII-35. The mass spectrum was obtained by scanning the RF and DC voltages exponentially with respect to time, but nevertheless maintaining the ratio of their amplitudes constant.

The process of ion acceleration was as follows. Positive ions were accelerated to the top surface of the quadrupole vacuum envelope biased at -8 V with respect to the rocket skin. Some of the ions entering through the orifice were further accelerated into the quadrupole structure and rod system which were biased at -128 V. The RF and DC voltages were then adjusted so that only ions of a particular charge-to-mass ratio were able to traverse through the length of the rods to the collector system. All other ions hit the rods or envelopes and were scattered away from the collector.

The minimum current that could be detected by the electrometer system of the detector was 5×10^{-16} A and the electrometer amplifier had a response time of 20 msec for current pulses at 10^{-12} A. The RF sweep voltage was monitored on a continuous channel to identify the mass number being detected with the help of laboratory calibrations using a conventional ion source and known gases. The quadrupole vacuum pressure, measured with the help of a Pirani-type pressure gauge, was also continuously monitored. It was concluded that for all practical considerations, the sensitivity could be assumed to be invariant over the mass range. Any fluctuations observed were caused by instrumentation difficulties.

Significantly high background currents, which were directly

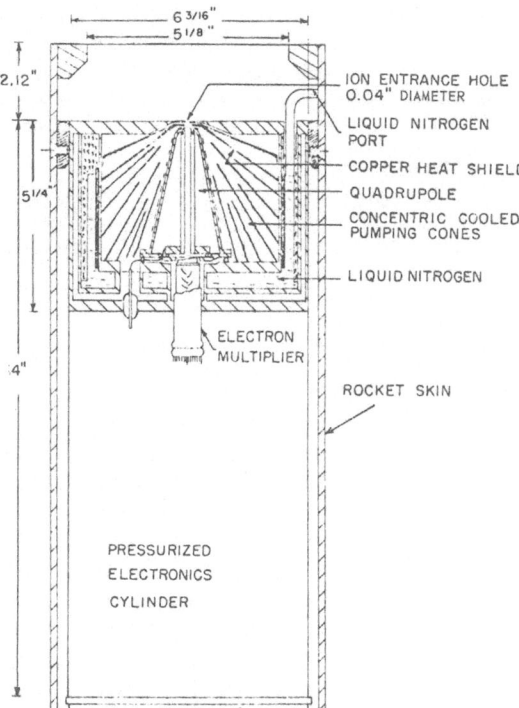

After R. S. Narcisi and A. D. Bailey, *J. Geophys. Res.* **70**, 3687, (1965).

Fig. VII-34. Cross section of the front end of the payload of the ion quadrupole mass spectrometer (D-region measurements).

proportional to the total ion current, were observed in the lower mass ranges, less than or equal to 4 amu. These background currents were attributed to the high value of the ion injection voltage. In the higher mass ranges, the background current was observed to be apparently independent of mass and a negligible portion of the total ion current.

In order to maintain the spectrometer within its operating pressure range, thereby making possible the measurements in the *D* region where the ambient gas pressure is relatively high, a liquid-nitrogen chilled, Zeolite adsorption pump which could be utilized in rocket flight was used. In actual flight, above 63 km ascending and down to 55 km descending, the pump maintained a steady pressure of 1μ in the system. Below 55 km the pressure was observed to increase slightly. It was found that at least three

**Table VII-7. Design Parameters for
Mass Spectrometer Constructed of
Stainless Steel Rods**

Mass range	1 to 46 amu
Mass resolution	16
Peak RF voltage	310 V
Ion injection voltage	128 V
RF frequency	6.0 Mc
Field radius	0.164 cm
Rod radius	0.1905 cm
Rod length	7.6248 cm
Injection port diameter	Not limited

hours of chilling with liquid nitrogen in the pump is required to ensure optimum performance of the pump.

The exact determination of mass number was further curtailed by errors in making the exact peak current determinations. A ringing effect was observed when the peak current approached 10^{-10} A and has been attributed to overcompensation in the feedback circuit of the electrometer amplifier. As a result of this effect, peak currents could only be read to an accuracy of approximately $\pm 40\%$. There were also statistical errors involved when reading smaller currents. Because of the lack of data on measurements made in the D region, the investigators have been cautious enough to assign an overall error of a factor of two to the measured current values.

An estimate of the sensitivity of the instrument was made by utilizing the positive-ion density measurements made by the electrostatic probes on the same flight and assuming that the total ion current in the mass spectra is proportional to the positive ion concentration outside of the instrument. The plot of sensitivity *versus* altitude is shown in Fig. VII-36.

It can be readily seen that the sensitivity of the instrument changed by two orders of magnitude from 67 to 80 km. By multiplying the sensitivity factor and the ion current, an approximate ion-density profile could be obtained. By observing the mass spectra obtained, it was noticed that there was a sharp transition in the nature of the mass spectra at 82.5 km.

a. Results for Measurements in the Range 64 to 82.5 km

The ions detected in this altitude range are shown in Fig. VII-37.

The dominant ions appeared to be 18^+, 30^+, and 37^+. Both 18^+ and 37^+ showed an extremely sharp decrease in density at 82.5 km while 30^+ continuously increased off scale. 37^+ seemed to exist only

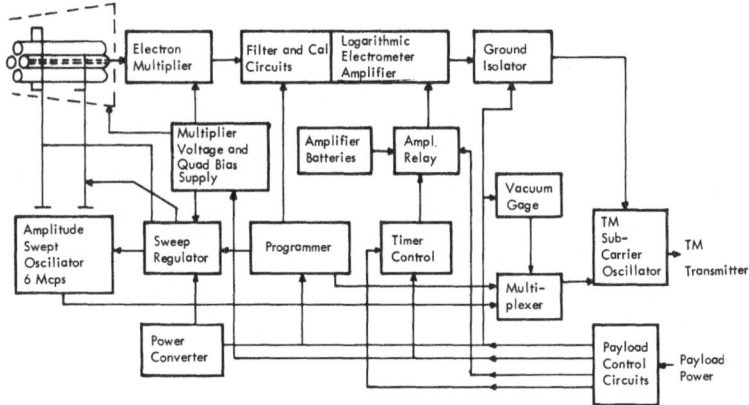

After R. S. Narcisi and A. D. Bailey, *J. Geophys. Res.* **70**, 3687, (1965).

Fig. VII-35. Electrical block diagram of the mass spectrometer.

below 86 km. 32^+, first appearing at 75 km, increased rapidly, approached the abundance of 30^+, and then proceeded off scale near 83 km. 17^+ seemed to decrease in density with increasing altitude similar to 18^+ and 37^+. It was observed that below 76 km approximately 50% of the total ion current is due to ions of mass greater than 45 amu. The detected ions were identified as follows: 17^+ as OH^+, 28^+ as N_2^+, 30^- as NO^+, 32^+ as O_2^+, and 37^+ as $H_5O_2^+$ (?), a cluster water-vapor ion.

Speculation about 37^+ ($H_5O_2^+$) seems to be well founded since Knewstubb and Tickner have observed ions of $H_3O^+ \cdot (H_2O)_n$ where $n = 0$ to 5 with a mass spectrometer in a DC glow discharge in water vapor. It was also speculated that heavy cluster ions like $H_5O_2^+$ could be responsible for more than half of the total ion current being concentrated in the higher mass ranges, that is, greater than 45 amu.

Hydroxyl (OH) light emission observed by nighttime rocket measurements of Tarasova seems to follow the peak current distribution pattern with altitude of the 17^+ ion below 83 km.

The mass spectrometer results of NO^+, O_2^+, and N_2^+ are compared to those calculated by Nicolet and Aiken in 1960 and represented in Fig. VII-38. Nicolet and Aiken, at that time, had considered only the effect of X-rays and Lyman-α radiation. Later, after including the effect of cosmic rays, the concentrations of N_2^+ and O_2^+ below 80 km showed an increase of more than an order of magnitude which was not observed by the mass spectrometer. Only a charge transfer reaction of the type $N_2 + O_2^+ \rightarrow NO^+ + NO$, as suggested by Nicolet and Swider in 1963, would be able to correct

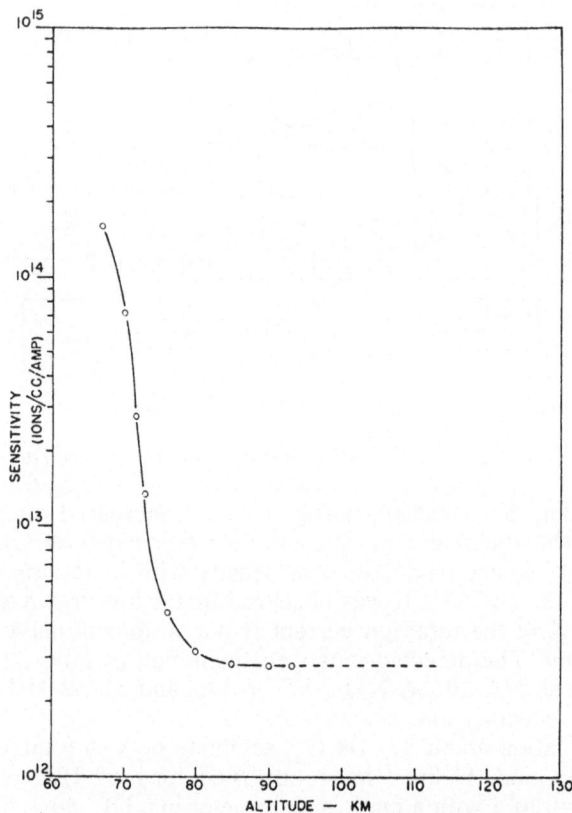

After R. S. Narcisi and A. D. Bailey, *J. Geophys. Res.* **70**, 3687, (1965).

Fig. VII-36. Sensitivity as a function of altitude to within a factor of four.

the discrepancies between theory and experiment. Also, the conclusion reached by Nicolet and Aiken that NO^+, O_2^+, and N_2^+ form the dominant ionic constituents of the D region is contradicted by the mass spectrometer results. These results showed that these ions form only about one-fifth of the total ionic content detected in the D region.

Water vapor ions in the D region could be caused by either X-ray and cosmic-ray ionization or charge-transfer reaction of water vapor with other ions. The charge-transfer mechanism could occur since such reactions were found to occur in plasma tunnel experiments when nitrogen and argon ions at $1000°$ K collided with background water vapor molecules in the tunnel. The sharp decrease

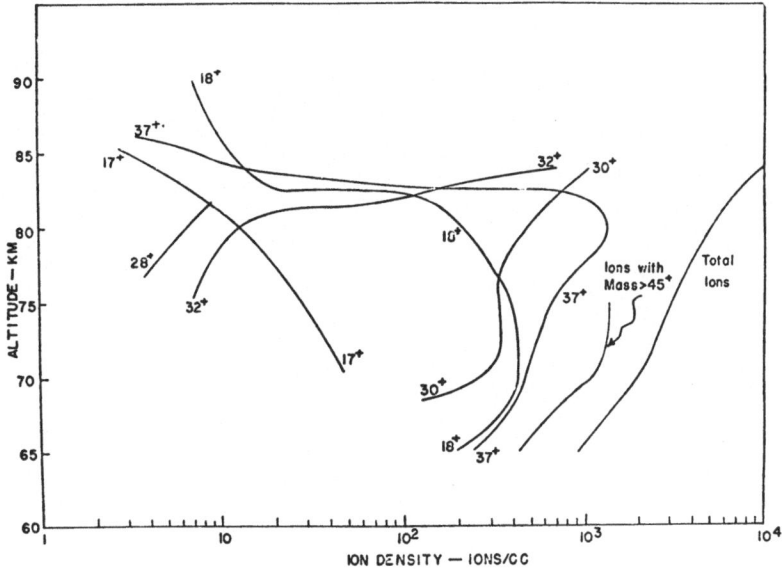

After R. S. Narcisi and A. D. Bailey, *J. Geophys. Res.* 70, 3687, (1965).

Fig. VII-37. Detected ions vs. altitude.

in the water vapor ion concentration at 82.5 km is supposed to be caused by the photodissociation of water vapor by radiations of wavelength less than 1800 Å in the reaction $H_2O + h\nu \rightarrow H + OH$.

It was proposed that 37^+ could not be caused by rocket contamination because of the disappearance of 37^+ above 86 km on ascent and its reappearance on descent below 86 km.

b. Results for Measurements in the Range 82.5 to 111.7 km

In this altitude range, ions 30^+ and 32^+ appeared to be the major constituents. Some of the new ion peaks observed were 23^+, 24^+, 25^+, 26^+, 40^+, and 42^+. Some of the calculated results were

$$\frac{n(24^+)}{n(25^+)} = \frac{n(24^+)}{n(26^+)} \approx 5.5 \pm 2.0 \tag{VII-44}$$

and

$$\frac{n(26^+)}{n(25^+)} \approx 1.06 \pm 0.25 \tag{VII-45}$$

Ions 23^+, 24^+, and 40^+ all seemed to show the same distinct altitude profile – a 10 km-wide peak with a maximum at 95 km and a mini-

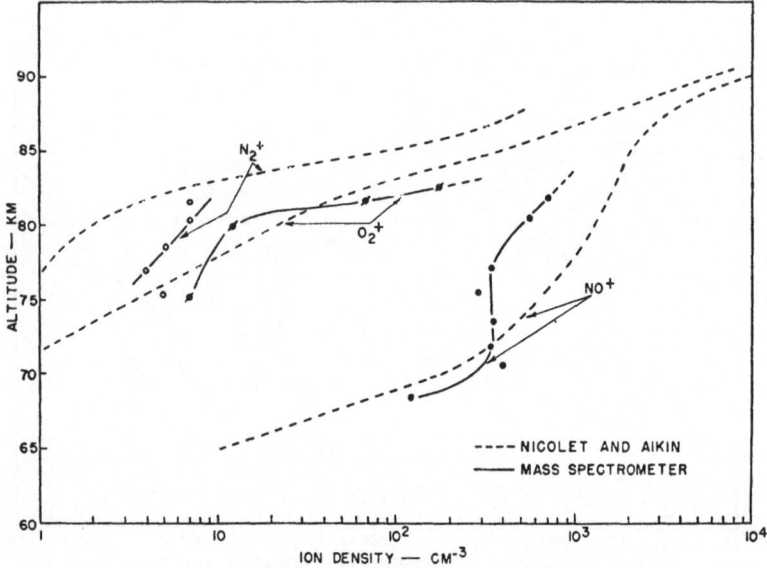

After R. S. Narcisi and A. D. Bailey, *J. Geophys. Res.* **70**, 3687, (1965).

Fig. VII-38. (NO^+), $O_2^+)$, and (N_2^+) experimental and theoretical vs. altitude.

mum at 105 km. Figure VII-39 represents the profile for the 23+ ion.

Ions 44+, 16+, and 15+ appeared to increase monotonically above 85 km; however, 16+ showed a peak at 106 km, as did 17+. Although very difficult to measure, the intensity of 28+ could be said to be less than the intensity of 24+. Figure VII-40 gives a composite picture of all the ions detected and theoretically estimated in this altitude range. The 30+ and 32+ ions which are major ions are supposed to be NO+ and O_2^+. Ions 23+, 24+, 25+, 26+, and 40+ were supposed to be metallic ions of possible meteoric origin and were identified as follows: Na+ as 23+, Mg+ as 24+, 25+, 26+, and Ca+ as 40+.

Istomin was the first to detect metallic ions in the upper atmosphere with a mass spectrometer as was discussed earlier. He detected magnesium, calcium, iron, and silicon in stratified layers at 105 and 120 km. However, he did not have enough data to support his discovery. Istomin detected these metallic ions, but could not determine their accurate altitude distributions. The experiments of Bullock and Hunten are extremely helpful in establishing the presence and distribution of sodium in the altitude range of interest.

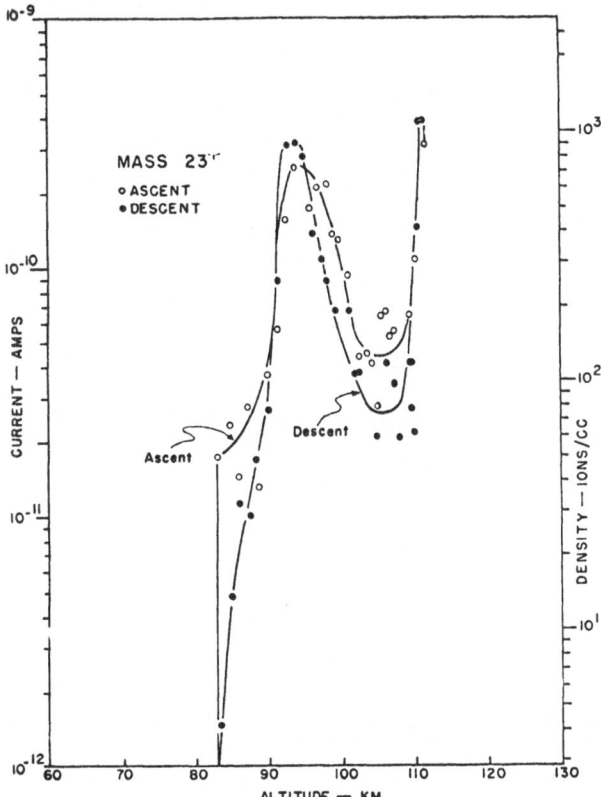

After R. S. Narcisi and A. D. Bailey, *J. Geophys. Res.* **70**, 3687, (1965).

Fig. VII-39. Profile for mass 23+ ion.

Their sodium atom profiles, obtained from the measurement of sodium light emissions in the twilight, appear exactly similar to the sodium ion profile below 105 km measured by Narcisi *et al.*

As corroborated by Hunten's estimate in 1954 that the ratio of neutral sodium to sodium ions ranges from 0.1 to 1 at altitudes near 100 km, the simple theory of Potter and Del Duca (1960) assumed that this ratio equals unity and approximately accounts for the observed profiles of the sodium atoms. Their theory assumed the following: (1) interaction of sodium with oxygen; (2) photoionization of sodium; (3) recombination of sodium ions with electrons and negative ions; and (4) photodissociation of oxides of sodium.

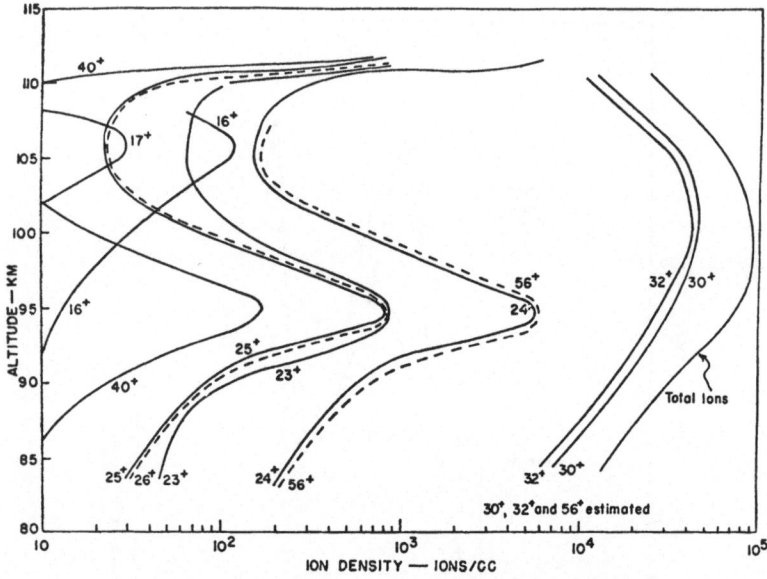

After R. S. Narcisi and A. D. Bailey, *J. Geophys. Res.* **70**, 3687, (1965).

Fig. VII-40. Composite picture of all the ions detected and theoretically estimated.

The measured concentration of magnesium isotope 24 seems to agree favorably with the generally accepted fact that it forms about 79% of the total concentration of all three isotopes (24, 25, and 26). Magnesium ions are also estimated to form about 10% of the total positive ions at 95 km. Istomin in 1963 identified mass 40^+ as calcium ions. He based his speculations on what was observed by Vallance-Jones in 1958 in the twilight spectra about ionized Ca II lines. Narcisi suggests that mass 40^+ could be an ion of magnesium oxide. According to the latter, the altitude distribution of the number of meteors reaching any one altitude is exactly similar to the distribution of metallic ions below 105 km.

Measurements of positive ion concentrations have given rise to a doubt that the nighttime E region might be maintained by ionized meteoric atoms. This idea is supported by the following two observations: (1) The observation of two distinct positive ion peaks at 100 and 120 km by Sagalyn and Smiddy in 1964 during a nighttime experiment and (2) the daytime metallic ion concentration in the E region equals the total ion density measured at night in the E region.

The validity of this doubt can only be decided after repeated experimentation and observation. Whether some of the minor constituents that were observed in the mass spectra were caused by rocket gas contamination or not is a problem not yet satisfactorily answered. Some of these ions are 44^+, 15^+, 17^+, and 18^+.

Narcisi *et al.* repeated their experiment on March 11, 1964, in the same location as before. It was estimated from this experiment that the concentration of iron ions is of the same order of magnitude as that of the magnesium ions. Istomin also obtained this result. It is interesting to note that the metallic ion profile measured in March is similar to the one measured in October, so that one can perhaps speak of a typical altitude distribution for the metallic ions.

The above experiments of Narcisi *et al.* indicate that with very careful instrumentation it is possible to explore the lower altitudes as thoroughly as possible. Not many details are presently known about the lower ionosphere and this can only be attributed to the lack of proper techniques. The instrumentation of Narcisi *et al.* plays an important role in further development and investigation of the physical processes of the D region.

12. A CONVENTIONAL MAGNETIC MASS SPECTROMETER DESIGNED FOR AN AERONOMY SATELLITE

Although the main purpose of this chapter is the discussion of the applications of nonmagnetic spectrometers in atmospheric composition research, it is highly interesting and rather justified to discuss a conventional double-focusing instrument designed for an aeronomy satellite. Such a mass spectrometer was developed by Spencer and Reber in 1961. The theory and operation of double-focusing instruments have been discussed in Chapter II. Any measurement which involves the sampling of particles of thermal or near-thermal energies by a satellite instrument involves an experimental situation wherein the measuring instrument disturbs the environment. To provide correction for this interference is a very difficult job but proper design will at least minimize the error.

The investigators therefore chose a design that would (1) minimize surface contacts of gases in order to reduce possible dissociation or recombination; (2) provide a relatively inaccessible hot filament to minimize the reaction effects of active gases with a hot surface; (3) provide for focusing of relatively high-energy ions, thus minimizing the effects of satellite velocity; and (4) provide an external ion-source geometry to reduce the problems caused by the consideration of gas flow into a sampling chamber. Other design characteristics which are essential for any space probe to provide efficient and steady performance are adequate sensitivity and electrical response, reasonably low weight and volume, a capacity to sustain reasonably high mechanical and thermal stress, and steady electrical characteristics in instrumentation.

The major elements of the system are the spectrometer tube, electrometer amplifier system, logic circuits, and an ionization electron-beam current regulator with associated power supplies. The instrument was so designed that the ion source projects outside the satellite surface. The ions pass through the lens assembly and converge on the slit just inside the satellite surface. Here the ions are deflected by an electric sector and assorted according to mass in the magnetic sector. They are then collected at various collectors, one for each mass. Because of the double focusing, the ion currents reaching the collectors are independent of first- and second-order energy variation and first-order ion-source entrance-angle variation. Minor but significant aspects of the design are: (1) The total ion-beam current can be measured periodically at the electric sector, thus providing a measure of the sum of the individual ion currents; (2) The ground-potential element shields the repeller and ion source from ionospheric electron flow; and (3) The entire ion source is covered by a vacuum-tight cap until orbit is achieved, at which time the cap is ejected.

The electrometer sensitivity is 10^{-15} A and the electrometer is switched from collector to collector with low and high sensitivity positions at each collector. Periodic calibration of the electrometer amplifier system is provided by a staircase voltage generated in the electrometer feedback loop. The mass spectrometer has a total weight of 12 lb and has a power requirement of 20 W. It is baked and sealed prior to exposure in orbit and has a sensitivity of approximately 10^{-11} torr. The atomic mass units 4, 14, 16, 18, 28, and 32 are those to which the instrument is tuned. This instrument is believed to be suitable for measurements from an earth satellite with an apogee of 700 to 1000 km. Recently Nier and others have been successfully flying magnetic mass spectrometers of the double-focusing type.

13. PARTICULAR EXPERIMENTAL AND ENGINEERING PROBLEMS

In this section some of the important and general problems of an experimental and engineering nature encountered in the utilization of mass spectrometers on rockets and satellites are discussed. Some problems have already been mentioned in the proper context; this section is only meant to provide a panoramic view of the most common ones.

a. Charge Acquired by the Vehicle

A negative vehicle potential caused by the negative charge acquired by the vehicle shifts the mass scale of the instrument toward the lighter masses. The effect of this potential will depend upon the field configuration around the satellite and upon the distance at which the ion is produced or at which it experiences its last collision. An increase in the velocity spread of the ions entering the analyzer is effected, thus reducing the resolution of the instrument. The negative potential may also lead to the appearance of false or harmonic peaks.

b. Desirability of a Proper Vehicle and Its Orientation in Space

Considering the information obtained in terms of altitude ranges and geographical area covered, one successful satellite launching should be worth hundreds of rocket shots. However. the satellite creates its own problems. The vehicle potential is dependent upon altitude, geographical coordinates, and the time of day. Therefore, an automatic adjustment of retarding potential in accordance with the accumulated charge is mandatory.

The orientation of the mouth of the spectrometer tube relative to the satellite velocity vector is an important factor in the operation of the RF mass spectrometer. If the mouth is aimed to the rear, the vacuum cone formed behind the satellite gives a false impression of the absence of ionization. With the mouth aimed forward, an assumed vehicle velocity is superimposed upon the ion velocity components directed along the tube axis. This assumed velocity will not cause an additional spread in the thermal velocities of the ions and will not affect the mass resolution adversely. It will merely shift the peaks of the ion currents toward the lower mass numbers on the mass scale. This shift is produced because the RF mass analyzer operates essentially as a velocity filter, in which only the ions that pass through at a certain fixed (synchronous) velocity reach the collector. If the ions have such an ordered velocity component prior to entering the instrument, less sweep voltage is required to acquire a synchronous velocity for all masses. Consequently, the corresponding ion-current peaks will appear shifted toward the lighter masses.

An additional velocity of $v = 8 \times 10^5$ cm/sec is equivalent to changing the accelerating saw-tooth voltage by $\Delta V_{eq} = mv^2/2e$. If

$m = 2$ amu (hydrogen), then $\Delta V_{eq} = 0.68$ V, and if $m = 40$ amu (argon), then $\Delta V_{eq} = 13.6$ V. If an instrument is designed for 5 V/amu, then the ion-current peaks will shift by 0.136 and 2.72 amu for hydrogen and argon, respectively.

If the mouth of the spectrometer tube is located perpendicular to the satellite velocity vector, then neither the resolution nor mass scale should change. All that can occur is an apparent reduction in the relative content of heavy ions. For a stabilized mass spectrometer in space, the best working arrangement is that the mouth of the tube be perpendicular to flight direction.

c. Engineering Problems

The size and weight of a mass spectrometer present a major problem. Another serious limitation in the use of a satellite-borne instrument is the capacity of the radio-telemetering memory system. The transmission of a mass spectrum that may comprise several tens of ion-current peaks necessitates telemetry channels of high resolving power. For example, in one of the experiments of Meadows and Townsend, the mass spectrum was transmitted over three channels at a sampling rate of approximately 300 per second and one high-capacity channel with a sampling rate of 1200 per second. The available telemetry systems are satisfactory for direct transmission from a rocket as far as the volume of transmitted information goes. However, it is not an easy problem to produce a memory system which has both (1) the capacity for the time required by the satellite to travel between two stations, and (2) the capability of transmitting all the stored information within the brief time interval required.

APPENDIX

Relative Concentrations and the Altitude Variations for Atomic Oxygen and Molecular Nitrogen

Pokhunkov, in 1962, discussed in detail the variation in the composition of the analyzed gases both in the entrance tube and in the analyzer of the mass spectrometer used in the atmospheric investigations. If the distance l from the entrance aperture to the ion source is significantly greater than the diameter d of the entrance aperture, only a negligible portion of the molecules in the stream entering the analyzer will reach the analyzer without colliding with the walls of the analyzer. Almost the entire entering stream and all of the emerging molecules will be ionized at the ion source only after multiple diffuse scattering at the walls and grids of the analyzer.

As a consequence of the rocket experiments in 1959 and 1960 in the 100 to 210 km range, Pokhunkov presented the results of two experiments performed with analyzers of different designs. In the 1959 experiment, Experiment 1, the peculiar design of the ion source and the long entrance tube prevented a direct infall of atmospheric atoms and molecules into the ionizing region. The ion current of atomic oxygen recorded by the analyzer may be represented by the sum of two currents as

$$I = I^i(\text{incident}) + I^r(\text{reflected}) \tag{A-1}$$

where I^i is incident current, I^r is reflected current, and $I^r < I^i$. Assuming that in an analyzer such as Experiment 1 there is no recombination in the entering stream, then

$$I^r < I^{r'} < I^i \tag{A-2}$$

where $I^{r'}$ is the ion current from ionization of the reflected stream in a hypothetical analyzer having no recombination in its entering stream. Therefore,

$$I^r + I^i \approx 2\,I^{r'} \tag{A-3}$$

The analyzer of the 1960 experiment, Experiment 2, was close to ideal. A short tube with a large diameter entrance aperture did minimize the recombination of atomic oxygen in the entering stream of this analyzer. Orientation of the entrance aperture of the analyzer relative to the flow produced no significant changes in the ratio of the O and N_2 ion currents as one would expect. Therefore, I^i, the ion current formed by ionization of the entering stream, may be regarded as corresponding to the O concentration in the undisturbed atmosphere. The correction factor for determining the undistorted atomic oxygen concentration is given by

$$K = \frac{2\,I^i}{I^r + I^i} \tag{A-4}$$

if (1) the relative concentration of atomic oxygen in the atmosphere was the same in both experiments and (2) the drop in the atomic oxygen concentration in the reflected stream was the same in both experiments. Therefore, for the analyzer in Experiment 1, $2\,I^{r'} = 1$, and for the analyzer in Experiment 2, $I^i + I^{r'} = 4$, so that, in this case, $K \approx 1.75$.

The first condition (1) for equation (A-4) presupposes that the ratio of the concentration of atomic oxygen to that of molecular nitrogen is the same in the two cases considered. This may not be

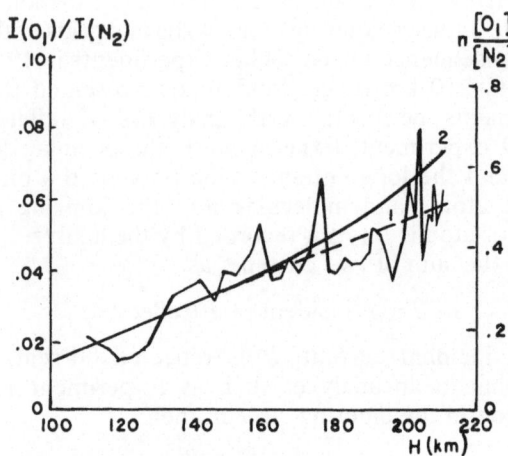

After A. A. Pokhunkov, from p. 145, *Artificial Earth Satellites – Vol. 12*, (Translated from Russian by Consultants Bureau Inc., New York, 1963).

Fig. A-1. Concentration and the ion current ratios of atomic oxygen to molecular nitrogen.

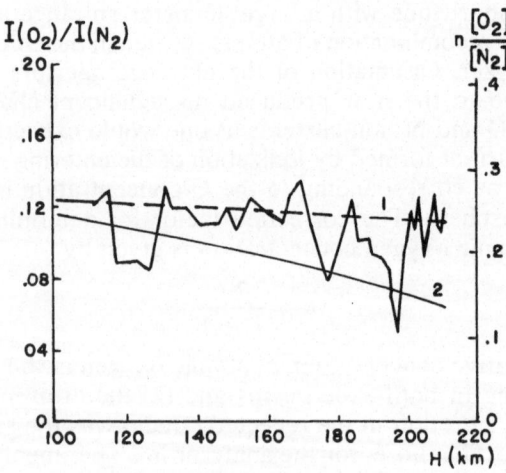

After A. A. Pokhunkov, from p. 145, *Artificial Earth Satellites – Vol. 12*, (Translated from Russian by Consultants Bureau, Inc., New York, 1963).

Fig. A-2. Concentration and the ion current ratios of molecular oxygen to molecular nitrogen.

true, but even a factor of two difference in the ratio would hardly change the value of K. The second condition (2) for equation (A-4) can have a significant effect on K. The analyzers of Experiments 1 and 2 are identical except that the former is twice as long. The probability of recombination of atomic oxygen in the reflected stream would quite naturally be greater for the longer analyzer. To get the atomic oxygen concentration in the atmosphere, one must multiply the concentration measured by the factor K. Figures A-1 and A-2 represent the concentration ratios and ion current ratios of atomic oxygen to molecular nitrogen and molecular oxygen to molecular nitrogen, respectively.

References

Alvarez, L. W., and R. Cornog, "He³ in Helium," *Phys. Rev.* **56,** 379 (1939).

Alvarez, L. W., and R. Cornog, "Helium and Hydrogen of Mass 3," *Phys. Rev.* **56,** 613 (1939).

Aston, F. W., "The Photometry of Mass Spectra and Atomic Weights of Krypton, Xenon and Mercury," *Proc. Roy. Soc.* **A126,** 511 (1930).

Bainbridge, K. T., and E. B. Jordan, "The Equivalence of Mass and Energy," *Phys. Rev.* **44,** 123 (1933).

Bainbridge, K. T., and E. B. Jordan, "Mass Spectrum Analysis," *Phys. Rev.* **50,** 282 (1936).

Barnard, G. P., "Recent Research With an Experimental Mass Spectrometer," *J. Electron.* **1,** 78 (1955).

Bauer, S. H., "Simple Mass Spectrometer," *J. Phys. Chem.* **39,** 959 (1935).

Bennett, W. H., "A Rapid Scanning rf Mass-Spectroscope," *Phys. Rev.* **79,** 222 (1950).

Bennett, W. H., "rf Mass Spectrometer," *J. Appl. Phys.* **21,** 143 (1950).

Best, N. R., "Matrix Telemetering System," *Electronics* **23,** 82 (1950).

Bleakney, W., and J. A. Hipple, Jr., "A Perfect e/m Filter as a Mass Spectrograph," *Phys. Rev.* **49,** 884 (1936).

Bleakney, W., and J. A. Hipple, Jr., "A New Mass Spectrometer With Improved Focussing Properties," *Phys. Rev.* **53,** 521 (1938).

Bloch, F., and C. D. Jefferies, "A Direct Determination of the Magnetic Moment of the Proton in Nuclear Magnetics," *Phys. Rev.* **80,** 305 (1950).

Boyd, R. L. F., "A Mass Spectrometric Probe Method for Study of Gas Discharges," *Nature* **165,** 142 (1950).

Boyd, R. L. F., and D. Morris, "An rf Probe for Mass Spectrometric Analysis of Ion Concentrations," *Proc. Phys. Soc. (London)* **68,** 1 (1955).

Boyd, R. L. F., "Some Techniques of Physical Measurements," *Proc. Roy. Soc. (London), Ser. A* **253,** 516 (1959).

Boyd, R. L. F., "Space Science," *Nature* **186,** 749 (1960).

Browne, J. C., "Spherical Electrostatic Analyzer for Measurement of Nuclear Energies," *Rev. Sci. Instr.* **22,** 952 (1951).

Brubaker, W. M., "The Quadrupole Mass Filter," *Neuvieme Colloq. Spectroscopicum Internat.*, Lyon, France (1961).

Brubaker, W. M., and J. Tuul, "Performance of a Satellite Quadrupole Mass Filter," *Rev. Sci. Instr.* **35,** 1007 (1964).

Bullock, W. R., and D. M. Hunten, "Vertical Distribution of Sodium in the Upper Atmosphere," *Can. J. Phys.* **39,** 976 (1961).

Byram, E. T., T. A. Chubb, and H. Friedman, "Dissociation of Oxygen in the Upper Atmosphere," *Phys. Rev.* **98,** 1594 (1955).

Cameron, A. E., and D. F. Eggers, Jr., "An Ion-Velocitron," *Rev. Sci. Instr.* **19,** 605 (1948).

Cartan, L., "A New Method of Focussing Beams of Positive Ions," *J. Phys. Radium* **8,** 111 (1937).

Chackett, K. F., *et al.,* "Variation in the Chemical Composition of the Stratosphere Air," *Nature* **168,** 358 (1951).

Christensen, F. E., "A Demonstration Mass-Spectrometer," *Am. J. Phys.* **19,** 59 (1951).

Cooke-Yarborough, E. H., and M. C. B. Russell, "A Simple Mass-Spectrometer for Analysis of Stable Tracer-Elements," *J. Sci. Instr.* **30,** 474 (1953).

Danilov, A. D., V. G. Istomin, and S. M. Poloskov, "Ionosphere Composition Investigated by Rockets and Satellites and Physical Processes Determining the Structure of the Ionosphere," *Space Research,* Vol. II (H. C. Van de Hulst, C. de Jager, and A. F. Moore, eds.), Amsterdam, North-Holland Publishing Co. (1961), p. 993.

Dekleva, J., "Improved Resolving Power of the rf Mass Spectrometer by Changing the Signal Shape," *Rev. Sci. Instr.* **26,** 399 (1955).

Dekleva, J., and M. Ribaric, "Some Design Data for a Non-Magnetic rf Mass-Spectrometer," *Rev. Sci. Instr.* **28,** 365 (1957).

Dempster, A. J., "New Method of Positive Ray Analysis," *Phys. Rev.* **3,** 316 (1918).

Dempster, A. J., "New Methods in Mass Spectroscopy," *Proc. Am. Phil. Soc.* **75,** 755 (1935).

Dempster, A. J., "Ion-Sources for Mass Spectroscopy," *Rev. Sci. Instr.* **7,** 46 (1936).

Dibeler, V. H., "Diaphragm Type Micromanometer for Use on a Mass Spectrometer," *J. Res. Natl. Bur. Std.* **46,** 1 (1951).

Dibeler, V. H., "Chemical Analysis," *Anal. Chem.* **26,** 58 (1954).

Farmer, J. B., *Mass Spectrometry* (C. A. McDowell, ed.), New York, McGraw-Hill Book Co. (1963), pp. 25–36.

Fisher, E., "The Three-Dimensional Stabilization of Charge Carriers in a [H. F. Electric] Quadrupole Field," *Z. Physik* **156**, 1 (1959).

Flesh, G. D., and H. J. Svec, "The Mass Spectra of Chromyl Chloride, Chromyl Chlorofluoride, Chromyl Fluoride," *J. Am. Chem. Soc.* **81**, 1787 (1959).

Flesh, G. D., and H. J. Svec, "Model of a Mass-Spectrometer Which Simultaneously Collects Positive and Negative Ions," *Rev. Sci. Instr.* **34**, 897 (1963).

Fox, R. E., "Ionization Potentials and Probabilities Using a Mass-Spectrometer," *Phys. Rev.* **84**, 859 (1951).

Friedman, H., "Rocket Observations of the Ionosphere," *Proc. IRE* **47**, 272 (1959).

Frost, D. C., *Mass Spectrometry* (C. A. McDowell, ed.), New York, McGraw-Hill Book Co. (1963), p. 92.

Gates, W. C., *Proc. Instr. Soc. Am.* **15**, 1 (1960).

Glenn, W. E., Jr., "A Time-of-Flight Mass Spectrograph," U.S. Atomic Energy Commission Report, AECD-3337 (1952).

Goodrich, G. W., and W. C. Wiley, "Resistance Strip Magnetic Electron Multiplier," *Rev. Sci. Instr.* **32**, 846 (1961).

Goodrich, G. W., and W. C. Wiley, "Continuous Channel Electron Multiplier," *Rev. Sci. Instr.* **33**, 761 (1962).

Goudsmit, S. A., "A Time-of-Flight Mass Spectrometer," *Phys. Rev.* **74**, 622 (1948).

Goudsmit, S. A., "A Method of Measuring Scattering of Particle Tracks," *Phys. Rev.* **74**, 1537 (1948).

Green, J. H., D. M. Pinkerton, and K. R. Ryan, *Advances in Mass Spectrometry*, Vol. 2 (R. M. Elliott, ed.), London, Pergamon Press, Inc. (1963), p. 477.

Hagelbarger, D. W., *et al.*, "Does Diffusive Separation Exist in the Atmosphere Below 55 Kilometers?" *Phys. Rev.* **82**, 107 (1951).

Harrower, G. A., *Astrophys. J.* **132**, 22 (1960).

Hays, E. E., P. I. Richards, and S. A. Goudsmit, "Magnetic Time-of-Flight Mass Spectrometer," *Phys. Rev.* **76**, 180 (1949).

Hays, E. E., P. I. Richards, and S. A. Goudsmit, "Mass Measurements with a Magnetic Time-of-Flight Mass-Spectrometer," *Phys. Rev.* **84**, 824 (1951).

Hays, E. E., "Goudsmit Type Time-of-Flight Mass-Spectrometer," *Phys. Rev.* **96**, 1454 (1954).

Henson, A. W., "A Modification of rf Mass Spectrometer," *J. Appl. Phys.* **21**, 1063 (1950).

Herzog, R., "Ionic- and Electro-Optical Cylindrical Lenses," *Z. Physik* **89**, 447 (1934).

Herzog, R., and J. Mattauch, "Theoretical Investigations of the Mass Spectrometer Without Magnetic Field," *Ann. Physik* **19**, 345 (1934).

Hintenberger, H. Von, and J. Mattauch, "Spurious Lines Observed in the Mass Spectrometer Without Magnetic Field," *Z. Physik* **106**, 279 (1937).

Hipple, J. A., H. Sommer, and H. A. Thomas, "A Precise Method of Determining the Faraday Constant by Magnetic Resonance," *Phys. Rev.* **76**, 1877 (1949).

Hipple, J. A., H. Sommer, and H. A. Thomas, "The Omegatron," *Phys. Rev.* **78**, 332 (1950).

Hipple, J. A., "A Combination of Crossed Field and Time-of-Flight Mass Spectrometer," *Phys. Rev.* **85**, 712 (1952).

Hipple, J. A., and H. Sommer, "Mass Spectroscopy in Physics Research," *Natl. Bur. Std. (U.S.), Circ.*, No. 522 (1953), p. 123.

Holmes, J. C., "Emission Current Regulator for Rocket-Borne rf Mass-Spectrometer," *Rev. Sci. Instr.* **28**, 290 (1957).

Holmes, J. C., and C. Y. Johnson, "Positive Ions in the Ionosphere," *Astronautics* **4**, 30 (1959).

Horowitz, R., and H. E. Lagow, "Upper-Air Pressure and Density Measurements from 90–220 kms with the Viking Rocket," *J. Geophys. Res.* **62**, 57 (1957).

Hull, A. W., "Electronic Devices as Aids to Research," *J. Phys.* **2**, 409 (1932); *Phys. Rev.* **22**, 279 (1923).

Ichimiya, T., *et al.*, "Measurement of Positive Ion-Density in the Ionosphere by Sounding Rocket," *Nature* **190**, 156 (1961).

Istomin, V. G., *Artificial Earth Satellites* **4**, 171 (1960); **7**, 64 (1961); **6**, 127 (1961); **11**, 94 (1961).

Istomin, V. G., *Geomagnetizm i Aeronomiya* **3**, 359 (1961).

Istomin, V. G., "Absolute Concentration of the Ion Components of the Earth's Atmosphere at Altitudes Between 100 and 200 kms," *Planetary Space Sci.* **11**, 169 (1963).

Istomin, V. G., "Ions of Extra-Terrestrial Origin in the Earth's Ionosphere," *Space Research*, Vol. III (W. Priester, ed.), Amsterdam, North-Holland Publishing Co. (1963), p. 209.

Istomin, V. G., and A. A. Pokhunkov, "Mass Spectrometer Measurements of Atmospheric Composition in the USSR," *Space Research,* Vol. III (W. Priester, ed.), Amsterdam, North-Holland Publishing Co. (1963), p. 117.

Ivanov-Kholodny, G. S., "Intensity of Sun's Short-Wave Radiations and Rates of Ionization and Recombination Processes in the Ionosphere," *Geomagnetizm i Aeronomiya* **2,** 377 (1962).

Jefferies, C. D., "A Direct Determination of the Magnetic Moment of the Protons in Units of the Nuclear Magnetron," *Phys. Rev.* **81,** 1040 (1951).

Johnson, C. Y., "Mass Determination of Ions Detected by Bennett Ion rf Mass-Spectrometer," *J. Appl. Phys.* **29,** 740 (1958).

Johnson, C. Y., "Errata – Mass Determination of Ions Detected by Bennett Ion rf Mass-Spectrometer," *J. Appl. Phys.* **29,** 1134 (1958).

Johnson, C. Y., "Aeronomic Parameters from Mass-Spectrometry," *Ann. Geophys.* **17,** 100 (1961).

Johnson, C. Y., "Two-Stage Single-Cycle rf Mass-Spectrometer," *Space Research,* Vol. III (W. Priester, ed.), Amsterdam, North-Holland Publishing Co. (1963), p. 1144.

Johnson, C. Y., and J. P. Heppner, "Night Time Measurement of Positive and Negative Ion Composition to 120 kms by Rocket-Borne Spectrometer," *J. Geophys. Res.* **60,** 533 (1955).

Johnson, C. Y., and J. C. Holmes, "Ionospheric Positive Ions," *Space Research,* Vol. I (H. Kallmann-Bijl, ed.), Amsterdam, North-Holland Publishing Co. (1960), p. 417.

Johnson, C. Y., and E. B. Meadows, "First Investigation of Ambient Positive Ion Composition to 219 Kilometers by Rocket-Borne Spectrometer," *J. Geophys. Res.* **60,** 193 (1955).

Johnson, C. Y., E. B. Meadows, and J. C. Holmes, "Ion-Composition of the Arctic Atmosphere," *J. Geophys. Res.* **63,** 443 (1958).

Johnson, C. Y., *et al.,* "Results Obtained with Rocket-Borne Ion Spectrometers," *J. Geophys. Res.* **14,** 475 (1958).

Katzenstein, H. S., and S. S. Friedland, "A New Time-of-Flight Mass-Spectrometer," *Rev. Sci. Instr.* **26,** 324 (1955).

Keller, R., "The Determination of Mass Spectra by the Measurements of Times of Flight," *Helv. Phys. Acta* **22,** 386 (1949).

Kerr, L. W., "Non-Magnetic Mass-Spectrometers," *J. Electron.* **2,** 179 (1958).

Kerwin, L., "Ion-Optics," *Mass Spectrometry* (C. A. McDowell, ed.), New York, McGraw-Hill Book Co. (1963).

Knewstubb, P. F., and A. W. Tickner, "Mass Spectrometry of Ions in Glow Discharges," *J. Chem. Phys.* **38**, 464 (1963).

Kobayaski, "Spherical Condenser Type Spectrometer," *J. Phys. Soc. Japan* **8**, 135 (1953).

Koomen, M., R. Scolnik, and R. Tousey, "Distribution of Night Airglow [OI] 5577 A Na D Layers Measured from a Rocket," *J. Geophys. Res.* **61**, 304 (1956).

Lagow, H. E., and R. Horowitz, "Comparison of High-Altitude Rocket and Satellite Density Measurements," *Phys. Fluids* **1**, 478 (1959).

Lunt, R. W., and A. H. Gregg, "Occurrence of Negative Ions in the Glow Discharge Through Oxygen and Other Gases," *Trans. Faraday Soc.* **36**, 1062 (1940).

Mattauch, J., and R. Herzog, "New Mass-Spectrograph," *Z. Physik* **89**, 786 (1934).

McQueen, J. H., "Isotopic Separation Due to Settling in the Atmosphere," *Phys. Rev.* **80**, 100 (1950).

Meadows, E. B., and J. W. Townsend, Jr., "Neutral Gas Composition of the Upper Atmosphere by a Rocket-Borne Mass Spectrometer," *J. Geophys. Res.* **61**, 576 (1956).

Meadows, E. B., and J. W. Townsend, Jr., "IGY Rocket Measurements of Arctic Atmospheric Composition above 100 km," *Space Research*, Vol. II (H. Kallmann-Bijl, ed.), Amsterdam, North-Holland Publishing Co. (1960), p. 175.

Mirtov, S. A., "Rocket Investigations of Atmospheric Composition at High Altitudes," *The Russian Literature of Satellites*, Part II, New York, International Physical Index Inc. (1958), pp. 67–85.

Mirtov, S. A., and V. G. Istomin, "Investigation of Ion-Content of the Ionized Layers of the Atmosphere," *The Russian Literature of Satellites*, Part II, New York, International Physical Index Inc. (1958), pp. 119–131.

Narcisi, R. S., *et al.*, "Zeolite Adsorption Pump for Rocket-Borne Mass Spectrometer," *Translations of the National Vacuum Symposium* (G. H. Bancroft, ed.), New York, The Macmillan Company (1962), pp. 232–236.

Narcisi, R. S., *et al.*, "Calibration of a Flyable Mass Spectrometer for N and O Atom Sensitivity," *Space Research*, Vol. III (W. Priester, ed.), Amsterdam, North-Holland Publishing Co. (1963). p. 1156.

Narcisi, R. S., and A. D. Bailey, "Mass Spectrometric Measurements of Positive Ions at Altitudes from 64 to 112 Kilometers," *J. Geophys. Res.* **70**, 2687 (1965).

Nicolet, M., "Collision Frequency of Electrons in the Ionosphere," *J. Atmospheric Terrest. Phys.* **3**, 200 (1953).

Nicolet, M., and A. C. Aikin, "Formation of D-Region of the Ionosphere," *J. Geophys. Res.* **65**, 1469 (1960).

Nicolet, M., and W. Swider, Jr., "Ionospheric Conditions," *Planetary Space Sci.* **11**, 1459 (1963).

Nier, A. O., "A Mass-Spectrometer for Routine Abundance Measurements," *Rev. Sci. Instr.* **11**, 212 (1940).

Nier, A. O., "Mass Spectrometer for Leak Detection," *J. Appl. Phys.* **18**, 30 (1947).

Nier, A. O., E. P. Ney, and M. G. Inghram, "A Null Method for the Comparison of Two Ion Currents in a Mass Spectrometer," *Rev. Sci. Instr.* **18**, 294 (1947).

Nier, A. O., E. P. Ney, and M. G. Inghram, "A Mass-Spectrometer for Isotope and Gas Analysis," *Rev. Sci. Instr.* **18**, 398 (1947).

Nonaka, I., *et al.,* "A New Method of Measuring Ion Current of Mass Spectrometers," *J. Phys. Soc. Japan* **6**, 466 (1951).

Paul, W., and H. Steinwedel, "A New Mass Spectrometer Without a Magnetic Field," *Z. Naturforsch.* **8A**, 448 (1953).

Paul, W., and M. Raether, "The Electrical Mass Filter," *Z. Physik* **140**, 262 (1955).

Paul, W., H. P. Reinhard, and U. Von Zahn, "The Electric Mass Filter as a Mass Spectrometer and Isotope Separator," *Z. Physik* **152**, 143 (1958).

Peterlin, A., "Resolving Power of the Three-Grid System of the Bennett Type rf Mass Spectrometer." *Rev. Sci. Instr.* **26**, 398 (1955).

Pokhunkov, A. A., "The Study of the Upper Atmosphere Neutral Composition at Altitudes above 100 km," *Space Research,* Vol. I (H. Kallmann-Bijl, ed.), Amsterdam, North-Holland Publishing Co. (1960), p. 101.

Pokhunkov, A. A., "Gravitational Separation, Composition and the Structural Parameters of the Atmosphere at Altitudes above 100 km," *Space Research,* Vol. III (W. Priester, ed.), Amsterdam, North-Holland Publishing Co. (1963), p. 132.

Pokhunkov, A. A., "Variations in the Mean Molecular Weight of Night Air at Heights from 100 to 210 km Shown by Mass Spectrometry," *Artificial Earth Satellites* **12**, 145 (1963).

Potter, A. E., and B. S. Del Duca, "Origin of the Sodium Airglow," *J. Geophys. Res.* **65**, 3915 (1960).

Purcell, J. D., "The Focussing of Charged Particles by a Spherical Condenser," *Phys. Rev.* **54**, 818 (1938).

Redhead, P. A., "A Linear rf Mass-Spectrometer," *Can. J. Phys.* **30**, 1 (1952).

Redhead, P. A., and C. R. Crowell, "Analysis of Linear rf Mass-Spectrometer," *J. Appl. Phys.* **24**, 331 (1953).

Ridenour, L. N., and C. S. Lampson, "Thermionic Control of an Ionization Gauge," *Rev. Sci. Instr.* **8**, 162 (1937).

Robertson, A. J. B., B. W. Viney, and M. Warrington, "The Production of Positive Ions by Field Ionization at the Surface of a Thin Wire," *Brit. J. Appl. Phys.* **14,** 278 (1963).

Rocket Panel, "Pressure Densities and Temperature in the Upper Atmosphere," *Phys. Rev.* **88,** 1027 (1952).

Sagalyn, R. C., and M. Smiddy, "Rocket Investigations of the Lower Ionosphere," *Space Research,* Vol. IV (P. Müller, ed.), Amsterdam, North-Holland Publishing Co. (1964).

Sayers, J., "Self-Contained Measuring Equipment for Electron Density and Ionic Mass Spectrum," *Proc. Roy. Soc. (London), Ser. A* **253,** 522 (1959).

Schaeffer, E. J., and M. Nichols, "Mass Spectrometer for Upper-Air Measurements," *ARS J.* **31,** 1773 (1961).

Schaeffer, E. J., "The Dissociation of Oxygen Measured by a Rocket-Borne Mass Spectrometer," *J. Geophys. Res.* **68,** 1175 (1963).

Schaeffer, E. J., and M. Nichols, "Upper Air Neutral Composition Measurements by a Mass Spectrometer." *J. Geophys. Res.* **69,** 4649 (1964).

Seddon, J. C., "Continuous Electron Density Measurements up to 200 kms," *J. Geophys. Res.* **59,** 513 (1954); "Propagation Measurements in the Ionosphere with the Aid of Rockets," *J. Geophys. Res.* **58,** 323 (1953).

Sheridan, W. F., "rf Mass-Spectrometer as a Rocket-Borne Instrument for Atmospheric Composition Studies," *Phys. Rev.* **95,** 298 (1954).

Sloan, D. H., and W. M. Coates, "Recent Advances in the Production of Heavy High Speed Ions Without the Use of High Voltages," *Phys. Rev.* **46,** 539 (1934).

Smith, L. G., "Magnetic Electron Multipliers for Detection of Positive Ions," *Rev. Sci. Instr.* **22,** 166 (1951).

Smith, L. G., "A New Magnetic Period Mass-Spectrometer," *Rev. Sci. Instr.* **22,** 115 (1951).

Smith, L. G., "Mass Synchrometer," *Phys. Rev.* **91,** 481 (1953).

Smith, L. G., and C. C. Damm, "Mass Synchrometer Doublet Measurements at Masses 28 and 30," *Phys. Rev.* **90,** 324 (1953).

Smith, L. G., and C. C. Damm, "Mass Synchrometer," *Rev. Sci. Instr.* **27,** 638 (1956).

Smith, M. L., "Electromagnetically Enriched Isotopes and Mass Spectrometry," *Harwell Conf.,* September 1955.

Smythe, W. R., "A Velocity Filter for Electrons and Ions," *Phys. Rev.* **28,** 1275 (1926).

Smythe, W. R., and J. Mattauch, "A New Mass Spectrometer," *Phys. Rev.* **40**, 429 (1932).

Sommer, H., H. A. Thomas, and J. A. Hipple, "The Measurement of e/m by Cyclotron Resonance," *Phys. Rev.* **82**, 697 (1951).

Spencer, N. W., and C. A. Reber, "A Mass-Spectrometer for an Aeronomy Satellite," *Space Research,* Vol. III (W. Priester, ed.), Amsterdam, North-Holland Publishing Co. (1963), p. 1151.

Spencer-Smith, J., "Negative Ions of Iodine – Part II – Ion Beams," *Phil. Mag.* **19**, 1016 (1935).

Stephens, W. E., "A Pulsed Mass Spectrometer with Time Dispersion," *Phys. Rev.* **69**, 691 (1946).

Straus, H. A., "A New Mass-Spectrograph Isotopic Constitution of Nickel," *Phys. Rev.* **59**, 430 (1941).

Tarasova, T. M., "Night Sky Emission Line Intensity Distribution with Respect to Height," *Space Research,* Vol. III (W. Priester, ed.), Amsterdam, North-Holland Publishing Co. (1963), p. 166.

Thomas, H. A., and H. F. Finch, *Phys. Rev.* **78**, 787 (1950).

Townsend, J. W., Jr., "rf Mass-Spectrometer for Upper Atmospheric Research," *Rev. Sci. Instr.* **23**, 538 (1952).

Vallance-Jones, A., "Calcium and Oxygen in the Twilight Airglow," *Ann. Geophys.* **14**, 179 (1958).

Voorhies, H. G., C. F. Robinson, and L. G. Hall, *Advances in Mass Spectrometry,* Vol. I (J. D. Waldron, ed.), London, Pergamon Press, Inc. (1959), pp. 44 – 65.

Vorsin, A. N., *Radiochastotnyi Mass-Spektrometer,* Moscow, Academy of Sciences of USSR (1959).

Warren, J. W., *Nature* **165**, 810 (1950).

Warren, R. E., *et al.,* "Electrostatic Analyzer for Selection of Homogeneous Ion-Beam," *Rev. Sci. Instr.* **18**, 559 (1947).

Weisz, P. B., "Particle Accelerators as Mass-Analyzers," *Phys. Rev.* **70**, 91 (1946).

Wherry, T. C., and F. W. Karasek, "Performance of the Non-Magnetic rf Spectrometer Tube," *J. Appl. Phys.* **26**, 682 (1955).

Wiechert, E., "Velocity and Magnetic Deflection of Cathode Rays," *Göttingen Nachrichten* **3**, 260 (1898); *Wied. Ann.* **69**, 739 (1899).

Wiley, W. C., and I. H. McLaren, "Time-of-Flight Mass Spectrometer with Improved Resolution," *Rev. Sci. Instr.* **26**, 1150 (1955).

Wiley, W. C., and I. H. McLaren, "Bendix Time-of-Flight Mass Spectrometer," *Science* **124**, 817 (1956).

Wolff, M. M., and W. E. Stephens, "A Pulsed Mass-Spectrometer with Time Dispersion," *Rev. Sci. Instr.* **24**, 616 (1953).

Author Index

Subject Index

A

Aberrations
  α, 20, 23
  β, 22, 23
  first-order, velocity, 32
  fringing field, 23
Accelerator
  linear, 86, 101
  particle, 215
Aeronomic parameters, 211
Analysis
  gas, 213
  isotope, 213
Analyzer, 16, 66, 203, 204
  Bennett type, 95, 108, 160
  δ-function type, 66
  deflecting field type, 101
  electrostatic, 29, 30, 32, 35, 207, 215
  magnetic, 30, 32, 35
  radio-frequency, 96, 107, 168
  retarding barrier type, 101
  spherical, 27, 28
Arctic region, 150
Argon, 174
Atmosphere, 204, 212
  Arctic, 150, 153, 155, 211
  Earth's, 136, 173, 175, 210
  turbulent, 178
  upper, 131, 135–137, 155, 160, 170, 178, 181, 196, 208, 212, 214
Auroral zone, 156

B

Backing plate, 187, 190
  entrance chamber, 189
  geometry, 187

Bennett system
  five-stage, 82
  single-stage, 73
  three-stage, 76–78, 80, 82
  two-stage, 75, 76, 82
Blanking pulse, 60, 61
Bucking voltage, 39, 40
Bunching, 50
n-Butane components, 53, 83, 85

C

Chemical analysis, 208
Coincident field
  cycloidal, 32
  sector, 32
Collection efficiency, 110, 111
Composition
  atmospheric, 167, 171, 177, 199, 213, 214
    Arctic, 212
  chemical, 208
  ion, 135, 166, 167, 172, 211
    measurements, 189
    negative, 211
    positive, 154, 211
  measurements, 182, 189
  neutral gas, 135, 164, 174, 175, 212, 214
  studies, 179
Concentration
  atomic oxygen, 172
  ion, 168
Constant energy case, 44, 45
Constant momentum case, 44, 45
Constituents, atmospheric, 155, 172, 173

221